# MORE PRAISE FOR *ON THE FUTURE*

"Rees is hardly the first to issue a stern warning about what lies ahead if complacency and consumerism rule, but his lucid, well-reasoned explanation of the stakes and inimitable prose lift this manifesto above the rest. An impassioned call to action from one of the world's foremost scientists. A book to be read by anyone on Earth who cares about its future."

—*KIRKUS*, starred review

"[*On the Future*] offers forecasts of impending technological developments and words of hope for the human ability to use science to repair a wounded planet and improve lives. . . . This far-ranging but easily understood collection of ideas shares and communicates the enthusiasm of Rees's 'techno-optimist' view of the prospects for humanity."

—*PUBLISHERS WEEKLY*

"*On the Future* is a very important book that should be widely read and acted upon. Martin Rees combines his deep scientific insights and compassion for humanity's welfare to address, in clear and elegant prose, the major issues facing human civilization today, some of which are not now commonly considered. Whether or not you agree with all the points he makes, you must take them very seriously indeed."

—ROGER PENROSE, author of *Fashion, Faith, and Fantasy in the New Physics of the Universe*

"An engaging analysis of the most important issues facing the world, sprinkled with insight and suffused with wisdom and humanity."

—STEVEN PINKER, author of *Enlightenment Now*

"Are we heading for a utopian or dystopian future? Martin Rees believes it's down to us. But the one thing we must not do is put the brakes on technology. Science, applied wisely, offers humanity a bright future, but we must act now. In this visionary book, and despite his many fears, Rees adopts a refreshing and cautiously optimistic tone."

—**JIM AL-KHALILI**, author of *Paradox: The Nine Greatest Enigmas in Physics*

"A breathtaking journey through thrilling advances in science and technology that may address society's most vexing challenges, *On the Future* is ideal reading for all citizens of the twenty-first century."

—**MARCIA K. McNUTT**, president of the National Academy of Sciences

"What if we got one of the smartest people alive to figure the odds on how we might be able to survive our ability to do ourselves in? We have that person in Martin Rees, and his thoughtful answers in this book."

—**ALAN ALDA**

"As Yogi Berra said, 'It's tough to make predictions, especially about the future.' But in this readable and thought-provoking book, Martin Rees shows the challenges we and our planet face—and why scientists need to engage citizens in the choices that are made."

—**SHIRLEY M. MALCOM**, director of education and human resources programs at the American Association for the Advancement of Science

"For anyone who wants to consider the choices we have in our future and the implications of those choices, this is the book to read. Rees is a clear thinker and graceful writer, and he expresses an encouraging optimism about the future, if we can avoid some of our current species-limiting behaviors. Rees's projections are grounded in today's scientific knowledge and a scientist's sense of probability and presented with a deep sense of humility."

—**RUSH D. HOLT**, CEO of the American Association for the Advancement of Science and former U.S. Representative from New Jersey

"*On the Future* is a delightful intellectual treat and a must-read for everyone on Earth—and beyond, if aliens exist. With wisdom and originality, Martin Rees, our most accomplished living astronomer, addresses the most important of subjects—the future of humanity and the scientific advances and risks it brings. His deep personal insights are unique and exciting and his many anecdotes are enjoyable. I couldn't put it down."

—**ABRAHAM LOEB**, Harvard University

"In this outstanding book, the always brilliant Martin Rees addresses the key problems of our day, putting humanity's perils and prospects in perspective—from climate change, to the future of artificial intelligence, to the threat of bioterrorism, to the chance for future space adventurers to spread out into the universe. Rees is one of the deepest and clearest thinkers on these subjects, and his book sparkles with gems of insight and humor."

—**J. RICHARD GOTT**, coauthor of *Welcome to the Universe*

"In this short and mighty book, Rees grapples with the exhilarating promise and frightening possibilities of today's vast scientific advancements, situating himself somewhere between the techno-optimists and dystopian naysayers. *On the*

*Future* provides tremendous insight into science's great expanse and beseeches us all to get involved and advocate for long-term policies that will keep future generations safe. The future of humanity and the planet rests in our hands."

—**RACHEL BRONSON**, president and CEO of the *Bulletin of the Atomic Scientists*

"*On the Future* will captivate readers. Martin Rees stands at the top of our informed thinkers about futurology."

—**PEDRO G. FERREIRA**, author of *The Perfect Theory: A Century of Geniuses and the Battle over General Relativity*

"This inspiring and thought-provoking book by one of the world's leading scientists and visionaries is a must-read for anyone who cares about the future of humanity."

—**MAX TEGMARK**, author of *Life 3.0: Being Human in the Age of Artificial Intelligence*

"Our planet is in peril—and humanity needs huge doses of wisdom to save it. Fortunately, one man, Martin Rees, can provide it. This book is a must-read for all who care about our planet's future."

—**KISHORE MAHBUBANI**, author of *Has the West Lost It?*

"Martin Rees's book is a vital compass to help us navigate the future, an enthralling love letter to knowledge and rationality, and a call to arms for those of us who dare to hope for the best."

—**DAVID PUTTNAM**, film producer and educator

"Martin Rees's *On the Future* is a template of hope that offers practical scientific, social, and political solutions for avoiding man-made disasters that could devastate our species."

—**MICHAEL WILSON**, film producer

# ON THE FUTURE

# ON THE FUTURE

## PROSPECTS FOR HUMANITY

# MARTIN REES

PRINCETON UNIVERSITY PRESS    PRINCETON & OXFORD

Published by Princeton University Press
41 William Street, Princeton, New Jersey 08540
6 Oxford Street, Woodstock, Oxfordshire OX20 1TR

press.princeton.edu

Jacket image courtesy of Alamy Stock Photo

Extracts from *The Reith Lectures: Scientific Horizons* (first broadcast
on BBC Radio 4 in June 2010) are used with permission of BBC.

Library of Congress Control Number 2018935643
ISBN 978-0-691-18044-1

British Library Cataloging-in-Publication Data is available

This book has been composed in Adobe Text Pro and Refrigerator Deluxe

Printed on acid-free paper. ∞

Printed in the United States of America

10  9  8  7  6  5  4  3  2  1

# CONTENTS

# PREFACE

**THIS IS A BOOK ABOUT THE FUTURE.** I write from a personal perspective, and in three modes: as a scientist, as a citizen, and as a worried member of the human species. The book's unifying theme is that the flourishing of the world's growing population depends on the wisdom with which science and technology is deployed.

Today's young people can expect to live to the end of the century. So how can they ensure that ever more powerful technologies—bio, cyber, and AI—can open up a benign future, without threatening catastrophic downsides? The stakes are higher than ever before; what happens this century will resonate for thousands of years. In addressing such a wide-ranging theme I'm mindful that even the experts have a poor record of forecasting. But I'm unrepentant because it's crucial to enhance public and political discourse on long-term scientific and global trends.

The themes of this book have evolved and clarified through lectures for varied audiences, including the 2010 BBC Reith Lectures, published as *From Here to Infinity* (Martin Rees, *From Here to Infinity: Scientific Horizons* [London: Profile Books, 2011; New York: W. W. Norton, 2012]). I'm therefore grateful for the feedback from listeners and readers. And I acknowledge with special gratitude the input (knowing or unknowing) from friends and colleagues with specialised expertise, who are not specifically quoted in the text. Among them are (alphabetically) Partha Dasgupta, Stu Feldman, Ian Golden, Demis Hassabis, Hugh Hunt, Charlie Kennel, David King, Seán Ó hÉigeartaigh, Catharine Rhodes, Richard Roberts, Eric Schmidt, and Julius Weitzdorfer.

I am specially grateful to Ingrid Gnerlich of Princeton University Press for instigating the book, and for her advice while I was writing it. I'm also grateful to Dawn Hall for the copyediting, to Julie Shawvan for the index, to Chris Ferrante for the text design, and to Jill Harris, Sara Henning-Stout, Alison Kalett, Debra Liese, Donna Liese, Arthur Werneck, and Kimberley Williams from the Press for their efficiency in seeing the book through the publishing process.

# ON THE FUTURE

# INTRODUCTION

**A COSMIC CAMEO:**

Suppose aliens existed, and that some had been watching our planet for its entire forty-five million centuries, what would they have seen? Over most of that vast time-span, Earth's appearance altered very gradually. Continents drifted; ice cover waxed and waned; successive species emerged, evolved, and became extinct.

But in just a tiny sliver of Earth's history—the last hundred centuries—the patterns of vegetation altered much faster than before. This signalled the start of agriculture—and then urbanisation. The changes accelerated as human populations increased.

Then there were even faster changes. Within just fifty years the amount of carbon dioxide in the atmosphere began to rise abnormally fast. And something else unprecedented happened: rockets launched from the planet's surface escaped the biosphere completely. Some were propelled into orbits around the Earth; some journeyed to the Moon and other planets.

The hypothetical aliens would know that Earth would gradually heat up, facing doom in about six billion years when the Sun would flare up and die. But could they have predicted this sudden 'fever' halfway through its life—these human-induced alterations—seemingly occurring with runaway speed?

If they continued to keep watch, what would they witness in the next century? Will a final spasm be followed by silence? Or will the planet's ecology stabilise? And will an armada of rockets launched from Earth spawn new oases of life elsewhere?

This book offers some hopes, fears, and conjectures about what lies ahead. Surviving this century, and sustaining the longer-term future of our ever more vulnerable world, depends on accelerating some technologies, but responsibly restraining others. The challenges to governance are huge and daunting. I offer a personal perspective—writing partly as a scientist (an astronomer) but also as an anxious member of the human race.

* * *

For medieval Europeans, the entire cosmology—from creation to apocalypse—spanned only a few thousand years. We now envision time-spans a million times longer. But even in this vastly extended perspective, this century is special. It is the first when one species, ours, is so empowered and dominant that it has the planet's future in its hands. We've entered an era that some geologists call the Anthropocene.

The ancients were bewildered and helpless in the face of floods and pestilences—and prone to irrational dread. Large parts of the Earth were terra incognita. The ancients' cosmos was just the Sun and planets surrounded by the fixed stars spread across the 'vault of heaven'. Today, we know our Sun is one of one hundred billion stars in our galaxy, which is itself one of at least one hundred billion other galaxies.

But despite these hugely stretched conceptual horizons—and despite our enhanced understanding of the natural world, and control over it—the timescale on which we can sensibly plan, or make confident forecasts, has become shorter rather than longer. Europe's Middle Ages were turbulent and uncertain times. But these times played out against a 'backdrop' that changed little from one generation to the next; devotedly, medieval masons added bricks to cathedrals that would take a century to finish. But for us, unlike for them, the next century will be drastically different from the present. There has been an explosive disjunction between the ever-shortening timescales of social and technical change and the billion-year time-spans of biology, geology, and cosmology.

Humans are now so numerous and have such a heavy collective 'footprint' that they have the ability to transform, or even ravage, the entire biosphere. The world's growing and more demanding population puts the natural environment under strain; peoples' actions could trigger dangerous climate change and mass extinctions if 'tipping points' are crossed—outcomes that would bequeath a depleted and impoverished world to future generations. But to reduce these risks, we don't need to put the brakes on technology; on the contrary, we need to enhance our understanding of nature and deploy appropriate technology more urgently. These are the themes of chapter 1 of this book.

Most people in the world live better lives than their parents did—and the proportion in abject poverty has been falling. These improvements, against a backdrop of a fast-growing population, couldn't have happened without advances in science and technology—which have been positive forces in the world. I argue in chapter 2 that our lives, our health, and our environment can benefit still more from further progress in biotech, cybertech, robotics, and AI. To that extent, I am a techno-optimist. But there is a potential downside. These advances expose our ever more interconnected world to new vulnerabilities. Even within the next decade or two, technology will disrupt working patterns, national economies, and international relations. In an era when we are all becoming interconnected, when the disadvantaged are aware of their predicament, and when migration is easy, it is hard to be optimistic about a peaceful world if a chasm persists, as deep as it is in today's geopolitics, between welfare levels and life chances in different regions. It is specially disquieting if advances in genetics and medicine that can enhance human lives are available to only a privileged few and portend more fundamental forms of inequality.

There are some who promote a rosy view of the future, enthusing about improvements in our moral sensitivities as well as in our material progress. I don't share this perspective. There has plainly, thanks to technology, been a welcome improvement in most people's lives and life chances—in education, health, and lifespan. However, the gulf between the way the world is and the way it could be is wider than it ever was. The lives of medieval people may have been miserable, but there was little that could have been done to improve those lives. In contrast, the plight of the 'bottom billion' in today's world could be transformed by redistributing the wealth of the thousand richest people on the planet. Failure to respond to this humanitarian imperative, which nations have the power to remedy, surely casts doubt on any claims of institutional moral progress.

The potentials of biotech and the cyberworld are exhilarating—but they're frightening too. We are already, individually and collectively, so greatly empowered by accelerating innovation that we can—by design, or as unintended consequences—engender global changes that will resonate for centuries. The smartphone, the web, and their ancillaries are already crucial to our networked lives. But these technologies would have seemed magical even just twenty years ago. So, looking several decades ahead we must keep our minds open, or at least ajar, to transformative advances that may today seem like science fiction.

We can't confidently forecast lifestyles, attitudes, social structures, or population sizes even a few decades hence—still less the geopolitical context against which these trends will play out. Moreover, we should be mindful of an unprecedented kind of change that could emerge within a few decades. Human beings themselves—their mentality and their physique—may become malleable through the deployment of genetic modification and cyborg technologies. This is a game changer. When we admire the literature and artefacts that have survived from antiquity, we feel an affinity, across a time gulf of thousands of years, with those ancient artists and their civilisations. But we can have zero confidence that the dominant intelligences a few centuries hence will have any emotional resonance with us—even though they may have an algorithmic understanding of how we behaved.

The twenty-first century is special for another reason: it is the first in which humans may develop habitats beyond the Earth. The pioneer 'settlers' on an alien world will need to adapt to a hostile environment—and they will be beyond the reach of terrestrial regulators. These adventurers could spearhead the transition from organic to electronic intelligence. This new incarnation of 'life', not requiring a planetary surface or atmosphere, could spread far beyond our solar system. Interstellar travel is not daunting to near-immortal electronic entities. If life is now unique to the Earth, this diaspora will be an event of cosmic significance. Yet if intelligence already pervades the cosmos, our progeny will merge with it. This would play out over astronomical timescales—not 'mere' centuries. Chapter 3 presents a perspective on these longer-term scenarios: whether robots will supersede 'organic' intelligence, and whether such intelligence already exists elsewhere in the cosmos.

What happens to our progeny, here on Earth and perhaps far beyond, will depend on technologies that we can barely conceive today. In future centuries (still an instant in the cosmic perspective), our creative intelligence could jump-start the transitions from an Earth-based to a space-faring species, and from biological to electronic intelligence—transitions that could inaugurate billions of years of posthuman evolution. On the other hand, as discussed in chapters 1 and 2, humans could trigger bio, cyber, or environmental catastrophes that foreclose all such potentialities.

Chapter 4 offers some (perhaps self-indulgent) excursions into scientific themes—fundamental and philosophical—that raise questions about the extent of physical reality, and whether there are intrinsic limits to how much we'll ever understand of the real world's complexities. We need to assess what's credible, and what can be dismissed as science fiction, in order to forecast the impact of science on humanity's long-term prospects.

In the final chapter I address issues closer to the here and now. Science, optimally applied, could offer a bright future for the nine or ten billion people who will inhabit the Earth in 2050. But how can we maximise the chance of achieving this benign future while avoiding the dystopian downsides? Our civilisation is moulded by innovations that stem from scientific advances and the consequent deepening understanding of nature. Scientists will need to engage with the wider public and use their expertise beneficially, especially when the stakes will be immensely high. Finally, I address today's global challenges—emphasising that these may require new international institutions, informed and enabled by well-directed science, but also responsive to public opinion on politics and ethics.

Our planet, this 'pale blue dot' in the cosmos, is a special place. It may be a unique place. And we are its stewards in an especially crucial era. That is an important message for all of us—and the theme of this book.

# 1

# DEEP IN THE ANTHROPOCENE

## 1.1. PERILS AND PROSPECTS

A few years ago, I met a well-known tycoon from India. Knowing I had the English title of 'Astronomer Royal', he asked, 'Do you do the Queen's horoscopes'? I responded, with a straight face: 'If she wanted one, I'm the person she'd ask'. He seemed eager to hear my predictions. I told him that stocks would fluctuate, there would be new tensions in the Middle East, and so forth. He paid rapt attention to these 'insights'. But then I came clean. I said I was just an astronomer—not an astrologer. He abruptly lost all interest in my predictions. And rightly so: scientists are rotten forecasters—almost as bad as economists. For instance, in the 1950s an earlier Astronomer Royal said that space travel was 'utter bilge'.

Nor do politicians and lawyers have a sure touch. One rather surprising futurologist was F. E. Smith, Earl of Birkenhead, crony of Churchill and the UK's Lord Chancellor in the 1920s. In 1930 he wrote a book titled *The World in 2030*.[1] He'd read the futurologists of his era; he envisaged babies incubated in flasks, flying cars, and such fantasies. In contrast, he foresaw social stagnation. Here's a quote: 'In 2030 women will still, by their wit and charms, inspire the most able men towards heights that they could never themselves achieve'.

Enough said!

*   *   *

Back in 2003 I wrote a book which I titled *Our Final Century?* My UK publisher deleted the question mark. The American publishers changed the title to *Our Final Hour*.[2] My theme was this: Our Earth is forty-five million centuries old. But this century is the first in which one species—ours—can determine the biosphere's fate. I didn't think we'd wipe ourselves out. But I did think we'd be lucky to avoid devastating breakdowns. That's because of unsustainable stresses on ecosystems; there are more of us (world population is higher) and we're all more demanding of resources. And—even more scary—technology empowers us more and more, and thereby exposes us to novel vulnerabilities.

I was inspired by, among others, a great sage of the early twentieth century. In 1902 the young H. G. Wells gave a celebrated lecture at the Royal Institution in London.[3] 'Humanity', he proclaimed,

> has come some way, and the distance we have travelled gives us some insight of the way we have to go. . . . It is possible to believe that all the past is but the beginning of a beginning, and that all that is and has been is but the twilight of the dawn. It is possible to believe that all that the human mind has accomplished is but the dream before the awakening; out of our lineage, minds will spring that will reach back to us in our littleness to know us better than we know ourselves. A day will come, one day in the unending succession of days, when beings, beings who are now latent in our thoughts and hidden in our loins, shall stand upon this earth as one stands upon a footstool, and shall laugh and reach out their hands amidst the stars.

His rather purple prose still resonates more than a hundred years later—he realised that we humans aren't the culmination of emergent life.

But Wells wasn't an optimist. He also highlighted the risk of global disaster:

> It is impossible to show why certain things should not utterly destroy and end the human story . . . and make all our efforts vain . . . something from space, or pestilence, or some great disease of the atmosphere, some trailing cometary poison, some great emanation of vapour from the interior of the Earth, or new animals to prey on us, or some drug or wrecking madness in the mind of man.

I quote Wells because he reflects the mix of optimism and anxiety—and of speculation and science—which I will try to convey in this book. Were he writing today he would be elated by our expanded vision of life and the cosmos, but he would be even more anxious about the perils we face. The stakes are indeed getting higher; new science offers huge opportunities, but its consequences could jeopardise our survival. Many are concerned that it is 'running away' so fast that neither politicians nor the lay public can assimilate or cope with it.

\* \* \*

You may guess that, being an astronomer, anxiety about asteroid collisions keeps me awake at night. Not so. Indeed, this is one of the few threats that we can quantify—and be confident is unlikely. Every ten million years or so, a body a few kilometres across will hit the Earth, causing global catastrophe—so there are a few chances in a million that such an impact occurs within a human lifetime. There are larger numbers of smaller asteroids that could cause regional or local devastation. The 1908 Tunguska event, which flattened hundreds of square kilometres of (fortunately unpopulated) forests in Siberia, released energy equivalent to several hundred Hiroshima bombs.

Can we be forewarned of these crash landings? The answer is yes. Plans are afoot to create a data set of the one million potential Earth-crossing asteroids larger than 50 metres and track their orbits precisely enough to identify those that might come dangerously close. With the forewarning of an impact, the most vulnerable areas could be evacuated. Even better news is that we could feasibly develop spacecraft that could protect us. A 'nudge', imparted in space several years before the threatened impact, would only need to change an asteroid's velocity by a few centimetres per second to deflect it from a collision course with the Earth.

If you calculate an insurance premium in the usual way, by multiplying probability by consequences, it turns out to be worth spending a few hundred million dollars a year to reduce the asteroid risk.

Other natural threats—earthquakes and volcanoes—are less predictable. So far there is no credible way to prevent them (or even predict them reliably). But there's one reassuring thing about these events, just as there is about asteroids: their rate isn't increasing. It's about the same for us as it was for Neanderthals—or indeed for dinosaurs. But the consequences of such events depend on the vulnerability and value of the infrastructure that's at risk, which is much greater in today's urbanised world. There are, moreover, cosmic phenomena to which the Neanderthals (and indeed all pre-nineteenth-century humans) would have been oblivious: giant flares from the Sun. These trigger magnetic storms that could disrupt electricity grids and electronic communications worldwide.

Despite these natural threats, the hazards that should make us most anxious are those that humans themselves engender. These now loom far larger, and they are becoming more probable, and potentially more catastrophic, with each decade that passes.

We've had one lucky escape already.

## 1.2. NUCLEAR THREATS

In the Cold War era—when armament levels escalated beyond all reason—the superpowers could have stumbled towards Armageddon through muddle and miscalculation. It was the era of 'fallout shelters'. During the Cuban missile crisis, my fellow students and I participated in vigils and demonstrations—our mood lightened only by the 'protest songs', such as Tom Lehrer's lyrics: 'We'll all go together when we go, all suffused with an incandescent glow'. But we would have been even more scared had we truly realised just how close we were to catastrophe. President Kennedy was later quoted as having said that the odds were 'somewhere between one out of three and even'. And only when he was long retired did Robert McNamara state frankly that 'we came within a hairbreadth of nuclear war without realizing it. It's no credit to us that we escaped—Khrushchev and Kennedy were lucky as well as wise'.

We now know more details of one of the tensest moments. Vasili Arkhipov, a highly respected and decorated officer in Russia's navy, was serving as number two on a submarine carrying nuclear missiles. When the United States attacked the submarine with depth charges, the captain inferred that war had broken out and wanted the crew to launch the missiles. Protocol required the top three officers on board to agree. Arkhipov held out against such action—and thereby avoided triggering a nuclear exchange that could have escalated catastrophically.

Post-Cuba assessments suggest that the annual risk of thermonuclear destruction during the Cold War was about ten thousand times higher than the mean death rate from asteroid impact. And indeed, there were other 'near misses' when catastrophe was only avoided by a thread. In 1983 Stanislav Petrov, a Russian Air Force officer, was monitoring a screen when an 'alert' indicated that five Minuteman intercontinental ballistic missiles had been launched by the United States towards the Soviet Union. Petrov's instructions, when this happened, were to alert his superior (who could, within minutes, trigger nuclear retaliation). He decided, on no more than a hunch, to ignore what he'd seen on the screen, guessing it was a malfunction in the early warning system. And so it was; the system had mistaken the reflection of the Sun's rays off the tops of clouds for a missile launch.

Many now assert that nuclear deterrence worked. In a sense, it did. But that doesn't mean it was a wise policy. If you play Russian roulette with one or two bullets in the cylinder, you are more likely to survive than not, but the stakes would need to be astonishingly high—or the value you place on your life inordinately low—for this to be a wise gamble. We were dragooned into just such a gamble throughout the Cold War era. It would be interesting to know what level of risk other leaders thought they were exposing us to, and what odds most European citizens would have accepted, if they'd been asked to give informed consent. For my part, I would not have chosen to risk a one in three—or even a one in six—chance of a catastrophe that would have killed hundreds of millions and shattered the historic fabric of all European cities, even if the alternative were certain Soviet dominance of Western Europe. And, of course, the devastating consequences of thermonuclear war would have spread far beyond the countries that faced a direct threat, especially if a 'nuclear winter' were triggered.

Nuclear annihilation still looms over us: the only consolation is that, thanks to arms control efforts between the superpowers, there are about five times fewer weapons than during the Cold War—Russia and the United States each have about seven thousand—and fewer are on 'hair trigger' alert. However, there are now nine nuclear powers, and a higher chance than ever before that smaller nuclear arsenals might be used regionally, or even by terrorists. Moreover, we can't rule out, later in the century, a geopolitical realignment leading to a standoff between new superpowers. A new generation may face its own 'Cuba'—and one that could be handled less well (or less luckily) than the 1962 crisis was. A near-existential nuclear threat is merely in abeyance.

Chapter 2 will address the twenty-first-century sciences—bio, cyber, and AI—and what they might portend. Their misuse looms as an increasing risk. The techniques and expertise for bio- or cyberattacks will be accessible to millions—they do not require large special-purpose facilities like nuclear weapons do. Cybersabotage efforts like 'Stuxnet' (which destroyed the centrifuges used in the Iranian nuclear weapons programme), and frequent hacking of financial institutions, have already bumped these concerns up the political agenda. A report from the Pentagon's Science Board claimed that the impact of cyberattack (shutting down, for instance, the US electricity grid) could be catastrophic enough to justify a nuclear response.[4]

But before that let's focus on the potential devastation that could be wrought by human-induced environmental degradation, and by climate change. These interlinked threats are long-term and insidious. They stem from humanity's ever-heavier collective 'footprint'. Unless future generations tread more softly (or unless population levels fall) our finite planet's ecology will be stressed beyond sustainable limits.

## 1.3. ECO-THREATS AND TIPPING POINTS

Fifty years ago, the world's population was about 3.5 billion. It is now estimated to be 7.6 billion. But the growth is slowing. Indeed, the number of births per year, worldwide, peaked a few years ago and is now decreasing. Nonetheless, the world's population is forecast to rise to around nine billion, or even higher, by 2050.[5] This is because most people in the developing world are still young and have not had children, and because they will live longer; the age histogram for the developing world will come to look more like it does for Europe. The largest current growth is in East Asia, where the world's human and financial resources will become concentrated—ending four centuries of North Atlantic hegemony.

Demographers predict continuing urbanisation, with 70 percent of people living in cities by 2050. Even by 2030, Lagos, São Paulo, and Delhi will have populations greater than thirty million. Preventing megacities from becoming turbulent dystopias will be a major challenge to governance.

Population growth is currently underdiscussed. This may be partly because doom-laden forecasts of mass starvation—in, for instance, Paul Ehrlich's 1968 book *The Population Bomb* and the pronouncements of the Club of Rome—have proved off the mark. Also, some deem population growth to be a taboo subject—tainted by association with eugenics in the 1920s and '30s, with Indian policies under Indira Gandhi, and more recently with China's hard-line one-child policy. As it turns out, food production and resource extraction have kept pace with rising population; famines still occur, but they are due to conflict or maldistribution, not overall scarcity.[6]

We can't specify an 'optimum population' for the world because we can't confidently conceive what people's lifestyles, diet, travel patterns, and energy needs will be beyond 2050. The world couldn't sustain anywhere near its present population if everyone lived as profligately—each using as much energy and eating as much beef—as the better-off Americans do today. On the other hand, twenty billion could live sustainably, with a tolerable (albeit ascetic) quality of life, if all adopted a vegan diet, travelled little, lived in small high-density apartments, and interacted via super-internet and virtual reality. This latter scenario is plainly improbable, and certainly not alluring. But the spread between these extremes highlights how naive it is to quote an unqualified headline figure for the world's 'carrying capacity'.

A world with nine billion people, a number that could be reached (or indeed somewhat exceeded) by 2050, needn't signal catastrophe. Modern agriculture—low-till, water-conserving, and perhaps involving genetically modified (GM) crops, together with better engineering to reduce waste, improve irrigation, and so forth—could plausibly feed that number. The buzz phrase is 'sustainable intensification'. But there will be constraints on energy—and in some regions severe pressure on water supplies. The quoted figures are remarkable. To grow one kilogram of wheat takes 1,500 litres of water and several megajoules of energy. But a kilogram of beef takes ten times as much water and twenty times as much energy. Food production uses 30 percent of the world's energy production and 70 percent of water withdrawals.

Agricultural techniques using GM organisms can be beneficial. To take one specific instance, the World Health Organization (WHO) estimates that 40 percent of children under the age of five in the developing world suffer from vitamin A deficiency; this is the leading cause of childhood blindness globally, affecting hundreds of thousands of children each year. So-called golden rice, first developed in the 1990s and subsequently improved, delivers beta-carotene, the precursor of vitamin A, and alleviates vitamin-A deficiency. Regrettably, campaigning organisations, Greenpeace in particular, have impeded the cultivation of golden rice. Of course, there is concern about 'tampering with nature', but in this instance, new techniques could have enhanced 'sustainable intensification'. Moreover, there are hopes that a more drastic modification of the rice genome (the so-called C4 pathway) could enhance the efficiency of photosynthesis, thus allowing faster and more intensive growth of the world's number one staple crop.

Two potential dietary innovations do not confront a high technical barrier: converting insects—highly nutritious and protein rich—into palatable food; and making artificial meat from vegetable protein. As for the latter, 'beef' burgers (made mainly of wheat, coconut, and potato) have been sold since 2015 by a California company called Impossible Foods. It will be a while, though, before these burgers will satisfy carnivorous gourmands for whom beetroot juice is a poor substitute for blood. But biochemists are on the case, exploring more sophisticated techniques. In principle, it is possible to 'grow' meat by taking a few cells from an animal and then stimulating growth with appropriate nutrients. Another method, called acellular agriculture, uses genetically modified bacteria, yeast, fungi, or algae to produce the proteins and fats that are found in (for instance) milk and eggs. There is a clear financial incentive as well as an ecological imperative to develop acceptable meat substitutes, so one can be optimistic of rapid progress.

We can be technological optimists regarding food—and health and education as well. But it's hard not to be a political pessimist. Enhancing the life chances of the world's poorest people by providing adequate nourishment, primary education, and other basics is a readily achievable goal; the impediments are mainly political.

If the benefits of innovation are to be spread worldwide, there will need to be lifestyle changes for us all. But these need not signal hardship. Indeed, all can, by 2050, have a quality of life that is at least as good as profligate Westerners enjoy today—provided that technology is developed appropriately, and deployed wisely. Gandhi proclaimed the mantra: 'There's enough for everyone's need but not for everyone's greed'. This need not be a call for austerity; rather, it calls for economic growth driven by innovations that are sparing of natural resources and energy.

The phrase 'sustainable development' gained currency in 1987, when the World Commission on Environment and Development, chaired by Gro Harlem Brundtland, prime minister of Norway, defined it as 'development that meets the needs of the present—especially the poor—without compromising the ability of future generations to meet their own needs'.[7] We all surely want to 'sign up' to reach this goal in the hope that by 2050 there will be a narrower gap between the lifestyle that privileged societies enjoy and that which is available to the rest of the world. This can't happen if developing countries mimic the path to industrialisation that Europe and North America followed. These countries need to leapfrog directly to a more efficient and less wasteful mode of life. The goal is not anti-technology. More technology will be needed, but channeled appropriately, so that it underpins the needed innovation. The more developed nations must make this transition too.

Information technology (IT) and social media are now globally pervasive. Rural farmers in Africa can access market information that prevents them from being ripped off by traders, and they can transfer funds electronically. But these same technologies mean that those in deprived parts of the world are aware of what they are missing. This awareness will trigger greater embitterment, motivating mass migration or conflict, if these contrasts are perceived to be excessive and unjust. It is not only a moral imperative, but a matter of self-interest too, for fortunate nations to promote greater equality—by direct financial aid (and by ceasing the current exploitative extraction of raw materials) and also by investing in infrastructure and manufacturing in countries where there are displaced refugees, so that the dispossessed are under less pressure to migrate to find work.

Yet long-term goals tend to slip down the political agenda, trumped by immediate problems—and a focus on the next election. The president of the European Commission, Jean-Claude Juncker, said, 'We all know what to do; we just don't know how to get re-elected after we've done it'.[8] He was referring to financial crises, but his remark is even more appropriate for environmental challenges (and it's playing out now with the discouragingly slow implementation of the UN's Sustainable Development Goals).

There is a depressing gap between what could be done and what actually happens. Offering more aid is not in itself enough. Stability, good governance, and effective infrastructure are needed if these benefits are to permeate the developing world. The Sudanese tycoon Mo Ibrahim, whose company led the penetration of mobile phones into Africa, in 2007 set up a prize of $5 million (plus $200,000 a year thereafter) to recognise exemplary and noncorrupt leaders of African countries—and the Mo Ibrahim Prize for Achievement in African Leadership has been awarded five times.

The relevant actions aren't necessarily best taken at the nation-state level. Some of course require multinational cooperation, but many effective reforms need implementation more locally. There are huge opportunities for enlightened cities to become pathfinders, spearheading the high-tech innovation that will be needed in the megacities of the developing world where the challenges are especially daunting.

Short-termism isn't just a feature of electoral politics. Private investors don't have a long enough horizon either. Property developers won't put up a new office building unless they get payback within (say) thirty years. Indeed, most high-rise buildings in cities have a 'designed lifetime' of only fifty years (a consolation for those of us who deplore their dominance of the skyline). Potential benefits and downsides beyond that time horizon are discounted away.

What about the more distant future? Population trends beyond 2050 are harder to predict. They will depend on what today's young people, and those as yet unborn, will decide about the number and spacing of their children. Enhanced education and empowerment of women—surely a priority in itself—could reduce fertility rates where they're now highest. But this demographic transition hasn't yet reached parts of India and sub-Saharan Africa.

The mean number of births per woman in some parts of Africa—Niger, or rural Ethiopia, for instance—is still more than seven. Although fertility is likely to decrease, it is possible, according to the United Nations, that Africa's population could double again to four billion between 2050 and 2100, thereby raising the global population to eleven billion. Nigeria alone would then have as large a population as Europe and North America combined, and almost half of all the world's children would be in Africa.

Optimists remind us that each extra mouth brings also two hands and a brain. Nonetheless, the greater the population becomes, the greater will be the pressures on resources, especially if the developing world narrows its gap with the developed world in its per capita consumption. And the harder it will be for Africa to escape the 'poverty trap'. Indeed, some have noted that African cultural preferences may lead to a persistence of large families as a matter of choice even when child mortality is low. If this happens, the freedom to choose your family size, proclaimed as one of the UN's fundamental rights, may come into question when the negative externalities of a rising world population are weighed in the balance.

We must hope that the global population declines rather than increases after 2050. Even though nine billion can be fed (with good governance and efficient agribusiness), and even if consumer items become cheaper to produce (via, for instance, 3D printing) and 'clean energy' becomes plentiful, food choices will be constrained and the quality of life will be reduced by overcrowding and reductions in green space.

## 1.4. STAYING WITHIN PLANETARY BOUNDARIES

We're deep into the Anthropocene. This term was popularised by Paul Crutzen, one of the scientists who determined that the ozone in the upper atmosphere was being depleted by CFCs—chemicals then used in aerosol cans and refrigerators. The 1987 Montreal Protocol led to the ban on these chemicals. This agreement seemed an encouraging precedent, but it worked because substitutes existed for CFCs that could be deployed without great economic costs. Sadly, it's not so easy to deal with the other (more important) anthropogenic global changes consequent on a rising population, changes that are more demanding of food, energy, and other resources. All these issues are widely discussed. What's depressing is the inaction—for politicians the immediate trumps the long term; the parochial trumps the global. We need to ask whether nations need to give more sovereignty to new organisations along the lines of the existing agencies under the auspices of the United Nations.

The pressures of rising populations and climate change will engender loss of biodiversity—an effect that would be aggravated if the extra land needed for food production or biofuels encroached on natural forests. Changes in climate and alterations to land use can, in combination, induce 'tipping points' that amplify each other and cause runaway and potentially irreversible change. If humanity's collective impact on nature pushes too hard against what the Stockholm environmentalist Johan Rockström calls 'planetary boundaries',[9] the resultant 'ecological shock' could irreversibly impoverish our biosphere.

Why does this matter so much? We are harmed if fish populations dwindle to extinction. There are plants in the rain forest that may be useful to us for medicinal purposes. But there is a spiritual value too, over and above the practical benefits of a diverse biosphere. In the words of the eminent ecologist E. O. Wilson,

> At the heart of the environmentalist worldview is the conviction that human physical and spiritual health depends on the planet Earth. . . . Natural ecosystems—forests, coral reefs, marine blue waters—maintain the world as we would wish it to be maintained. Our body and our mind evolved to live in this particular planetary environment and no other.[10]

Extinction rates are rising—we're destroying the book of life before we've read it. For instance, the populations of the 'charismatic' mammals have fallen, in some cases to levels that threaten species. Many of the six thousand species of frogs, toads, and salamanders are especially sensitive. And, to quote E. O. Wilson again, 'if human actions lead to mass extinctions, it's the sin that future generations will least forgive us for'.

Here, incidentally, the great religious faiths can be our allies. I'm on the council of the Pontifical Academy of Sciences (an ecumenical body; its seventy members represent all faiths or none). In 2014 the Cambridge economist Partha Dasgupta, along with Ram Ramanathan, a climate scientist from the Scripps Institute in California, organised a high-level conference on sustainability and climate held at the Vatican.[11] This offered a timely scientific impetus into the 2015 papal encyclical 'Laudato Si'. The Catholic Church transcends political divides; there's no gainsaying its global reach, its durability and long-term vision, or its focus on the world's poor. The Pope was given a standing ovation at the United Nations. His message resonated especially in Latin America, Africa, and East Asia.

The encyclical also offered a clear papal endorsement of the Franciscan view that humans have a duty to care for all of what Catholics believe is 'God's creation'—that the natural world has value in its own right, quite apart from its benefits to humans. This attitude resonates with the sentiments beautifully expressed more than a century ago by Alfred Russel Wallace, co-conceptualiser of evolution by natural selection:

I thought of the long ages of the past during which the successive generations of these things of beauty had run their course . . . with no intelligent eye to gaze upon their loveliness, to all appearances such a wanton waste of beauty. . . . This consideration must surely tell us that all living things were not made for man. . . . Their happiness and enjoyments, their loves and hates, their struggles for existence, their vigorous life and early death, would seem to be immediately related to their own well-being and perpetuation alone.[12]

The papal encyclical eased the path to agreement at the Paris climate conference in December 2015. It eloquently proclaimed that our responsibility—to our children, to the poorest, and to our stewardship of life's diversity—demands that we don't leave a depleted and hazardous world.

We all surely hold these sentiments. But our secular institutions—economic and political—don't plan far enough ahead. I'll return in my final chapters to address the daunting challenges to science and to governance that these threats pose.

Regulations can help. But regulations won't gain traction unless the public mind-set changes.

Attitudes in the West to, for instance, smoking and driving drunk have transformed in recent decades. We need a similar change in attitude so that manifestly excessive consumption and waste of materials and energy—4 × 4 SUVs (disparaged as Chelsea tractors in London, where they clog the streets in up-market districts), patio heaters, brightly illuminated houses, elaborate plastic wrappings, slavish following of fast-changing fashions, and the like—become perceived as 'tacky' rather than stylish. Indeed, a trend away from excessive consumption may happen without exterior pressure. For my generation, our living space (a student room and later something more spacious) was 'personalised' by books, CDs, and pictures. Now that books and music can be accessed online, we will perhaps become less sentimental about 'home'. We will become nomadic—especially as more business and socialising can be done online. Consumerism could be replaced by a 'sharing economy'. If this scenario transpires, it will be crucial that developing nations transition directly towards this lifestyle, bypassing the high-energy, high-consumption stage through which Europe and the United States have passed.

Effective campaigns need to be associated with a memorable logo. The BBC's 2017 TV series *Blue Planet II* showed an albatross returning from wandering thousands of miles foraging in the southern oceans—and regurgitating for its young not the craved-for nutritious fish, but bits of plastic. Such an image publicises and motivates the case for recycling plastics, which otherwise accumulate in the oceans (and the food chains of the creatures that live there). Likewise, the longtime iconic image (albeit somewhat misleading) showing a polar bear clinging to a melting ice floe is emblematic of the climate change crisis—my next topic.

## 1.5. CLIMATE CHANGE

The world will get more crowded. And there's a second prediction: it will gradually get warmer. Pressures on food supplies, and on the entire biosphere, will be aggravated by the consequent changes in global weather patterns. Climate change exemplifies the tension between the science, the public, and the politicians. In contrast to population issues, it is certainly not underdiscussed—despite the fact that in 2017 the Trump regime in the United States banned the terms 'global warming' and 'climate change' from public documents. But the implications of climate change are dismayingly under-acted-on.

One thing is not controversial. The concentration of $CO_2$ in the air is rising, mainly due to the burning of fossil fuels. The scientist Charles Keeling measured $CO_2$ levels using an instrument at the Mauna Loa Observatory in Hawai'i, which has been operating continuously since 1958 (following Keeling's death in 2005 the programme is being continued by his son, Ralph). And it is not controversial that this rise leads to a 'greenhouse effect'. The sunlight that heats the Earth is reemitted as infrared radiation. But just as the glass in a greenhouse traps the infrared radiation (though it lets the light in) the $CO_2$ likewise acts as a blanket that traps heat in the Earth's atmosphere, land masses, and oceans. This has been understood since the nineteenth century. A rise in $CO_2$ will induce a long-term warming trend, superimposed on all the other complicated effects that make climate fluctuate.

Doubling of $CO_2$, if all other aspects of the atmosphere were unchanged, would cause 1.2 degrees (centigrade) of warming, averaged over the Earth—this is a straightforward calculation. But what is less well understood are associated changes in water vapour, cloud cover, and ocean circulation. We don't know how important these feedback processes are. The fifth report from the Intergovernmental Panel on Climate Change (IPCC), published in 2013, presented a spread of projections, from which (despite the uncertainties) some things are clear. In particular, if annual $CO_2$ emissions continue to rise unchecked we risk triggering drastic climate change—leading to devastating scenarios resonating centuries ahead, including the initiation of irreversible melting of ice in Greenland and Antarctica, which would eventually raise sea levels by many metres. It's important to note that the 'headline figure' of a global temperature increase is just an average; what makes the effect more disruptive is that the rise is faster in some regions and can trigger drastic shifts in regional weather patterns.

The climate debate has been marred by too much blurring between science, politics, and commercial interests. Those who don't like the implications of the IPCC projections have rubbished the science rather than calling for better science. The debate would be more constructive if those who oppose current policies recognise the imperative to refine and firm up the predictions—not just globally but, even more important, for individual regions. Scientists in Cambridge and California[13] are pursuing a so-called Vital Signs project, which aims to use massive amounts of climatic and environmental data to find which local trends (droughts, heat waves, and such) are the most direct correlates of the mean temperature rise. This could offer politicians something more relevant and easier to appreciate than a mean global warming.

The build-up rate of $CO_2$ in the atmosphere will depend on future population trends and the extent of the world's continuing dependence on fossil fuels. But even for a specific scenario for $CO_2$ emission, we can't predict how fast the mean temperature will rise, because of the 'climate sensitivity factor' due to uncertain feedback. The consensus of the IPCC experts was that business as usual, with a rising population and continuing dependence on fossil fuels, has a 5 percent chance of triggering more than six degrees warming in the next century. If we think of current expenditure on cutting $CO_2$ emissions as an insurance policy, the main justification is to avoid the small chance of something really catastrophic (as a rise of six degrees would be) rather than the 50 percent chance of something seriously damaging but which could be adapted to.

The goal proclaimed at the Paris conference was to prevent the mean temperature rise from exceeding two degrees—and if possible to constrain it to 1.5 degrees. This is an appropriate goal if we are to reduce the risk of crossing dangerous 'tipping points'. But the question is: how to implement it? The amount of $CO_2$ that can be released without violating this limit is uncertain by a factor of two, simply because of the unknown climate sensitivity factor. The target is therefore an unsatisfactory one—and will obviously encourage fossil fuel interests to 'promote' scientific findings that predict low sensitivity.

Despite the uncertainties—both in the science and in population and economic projections—two messages are important:

1. Regional disruptions in weather patterns within the next twenty to thirty years will aggravate pressures on food and water, cause more 'extreme events', and engender migration.
2. Under 'business as usual' scenarios in which the world continues to depend on fossil fuels, we can't rule out, later in the century, really

catastrophic warming, and tipping points triggering long-term trends like the melting of Greenland's ice cap.

But even those who accept both these statements and agree that there's a significant risk of climate catastrophe a century hence, will differ in how urgently they advocate action today. Their assessment will depend on expectations of future growth and optimism about technological fixes. Above all, however, it depends on an ethical issue—the extent to which we should limit our own gratification for the benefit of future generations.

Bjørn Lomborg achieved prominence (along with 'bogyman status' among many climate scientists) through his book *The Skeptical Environmentalist*. He has convened a Copenhagen Consensus of economists to pronounce on global problems and policy.[14] These economists apply a standard discount rate, thereby in effect writing off what happens beyond 2050. There is indeed little risk of catastrophe within that time horizon, so unsurprisingly they downplay the priority of addressing climate change compared to other ways of helping the world's poor. But, as Nicholas Stern[15] and Martin Woltzman[16] would argue, if you apply a lower discount rate—and, in effect, don't discriminate on grounds of date of birth and care about those who'll live into the twenty-second century and beyond—then you may deem it worth making an investment now to protect those future generations against the worst-case scenario.

Consider this analogy. Suppose astronomers had tracked an asteroid and calculated that it would hit the Earth in 2100, not with certainty, but with (say) 10 percent probability. Would we relax, saying that it's a problem that can be set on one side for fifty years—people will then be richer, and it may turn out then that it's going to miss the Earth anyway? I don't think we would. There would be a consensus that we should start straight away and do our damnedest to find ways to deflect it or mitigate its effects. We'd realise that most of today's young children will still be alive in 2100, and we care about them.

(As a parenthesis, I'd note that there's one policy context when an essentially zero discount rate is applied: radioactive waste disposal, where the depositories deep underground, such as that being constructed at Onkalo in Finland, and proposed [but then aborted] under Yucca Mountain in the United States, are required to prevent leakage for ten thousand or even a million years—somewhat ironic when we can't plan the rest of energy policy even thirty years ahead.)

## 1.6. CLEAN ENERGY—AND A 'PLAN B'?

Why do governments respond with torpor to the climate threat? It is mainly because concerns about future generations (and about people in poorer parts of the world) tend to slip down the agenda. Indeed, the difficulty of impelling $CO_2$ reductions (by, for instance, a carbon tax) is that the impact of any action not only lies decades ahead but also is globally diffused. The pledges made at the 2015 Paris conference, with a commitment to renew and revise them every five years, are a positive step. But the issues that gained prominence during that conference will slip down the agenda again unless there's continuing public concern—unless the issues still show up in politicians' in-boxes and in the press.

In the 1960s the Stanford University psychologist Walter Mischel did some classic experiments. He offered children a choice: one marshmallow immediately, or two if they waited for fifteen minutes. He claimed that the children who chose to delay their gratification became happier and more successful adults.[17] This is an apt metaphor for the dilemma nations face today. If short-term payback—instant gratification—is prioritised, then the welfare of future generations is jeopardised. The planning horizon for infrastructure and environmental policies needs to stretch fifty or more years into the future. If you care about future generations, it isn't ethical to discount future benefits (and dis-benefits) at the same rate as you would if you were a property developer planning an office building. And this rate of discounting is a crucial factor in the climate-policy debate.

Many still hope that our civilisation can segue smoothly towards a low-carbon future. But politicians won't gain much resonance by advocating a bare-bones approach that entails unwelcome lifestyle changes—especially if the benefits are far away and decades into the future. Indeed, it is easier to gain support for adaptation to climate change rather than mitigation because the benefits of the former accrue locally. For instance, the government of Cuba, whose coastal areas are especially vulnerable to hurricanes and a rise in sea level, has formulated a carefully worked-out plan stretching for a century.[18]

Nonetheless, three measures that could mitigate climate change seem politically realistic—indeed, almost 'win-win'.

First, all countries could improve energy efficiency and thereby actually save money. There could be incentives to ensure 'greener' design of buildings. This is not just a matter of improved insulation—it requires re-thinking construction as well. To take one example, when a building is demolished, some of its elements—steel girders and plastic piping, for instance—will hardly have degraded and could be reused. Moreover, girders could be more cleverly designed at the outset so as to offer the same strength with less weight, thereby saving on steel production. This exemplifies a concept that is gaining traction: the circular economy—where the aim is to recycle as much material as possible.[19]

Often, technical advances make appliances more efficient. It would then make sense to scrap the old ones, but only if the efficiency gain is at least enough to compensate for the extra cost of manufacturing the updated version. Appliances and vehicles could be designed in a more modular way so that they could be readily upgraded by replacing parts rather than by being thrown away. Electric cars could be encouraged—and could be dominant by 2040. This transition would reduce pollution (and noise) in cities. But its effect on $CO_2$ levels depends, of course, on where the electricity comes from that charges the batteries.

Effective action needs a change in mind-set. We need to value long-lasting things—and urge producers and retailers to highlight durability. We need to repair and upgrade rather than replace. Or do without. Token reductions may make us feel virtuous but won't be enough—if everyone does a little, we'll only achieve a little.

A second 'win-win' policy would target cuts to methane, black carbon, and CFC emissions. These are subsidiary contributors to greenhouse warming. But unlike $CO_2$ they cause local pollution too—in Chinese cities, for instance—so there's a stronger incentive to reduce them. (In European countries the effort to reduce pollution starts off with a handicap. In the 1990s there was pressure in favour of diesel cars because of their greater fuel economy. This is only now being reversed because they emit polluting microparticles that endanger healthy living in cities.)

But the third measure is the most crucial. Nations should expand Research and Development

(R&D) into all forms of low-carbon energy generation (renewables, fourth-generation nuclear, fusion, and the rest), and into other technologies where parallel progress is crucial—especially storage and smart grids. That is why an encouraging outcome of the 2015 Paris conference was an initiative called Mission Innovation. It was launched by President Obama and by the Indian prime minister, Narendra Modi, and endorsed by the countries of the G7 plus India, China, and eleven other nations. It is hoped they'll pledge to double their publicly funded R&D into clean energy by 2020 and to coordinate efforts. This target is a modest one. Presently, only 2 percent of publicly funded R&D is devoted to these challenges. Why shouldn't the percentage be comparable to spending on medical or defence research? Bill Gates and other private philanthropists have pledged a parallel commitment.

The impediment to 'decarbonising' the global economy is that renewable energy is still expensive to generate. The faster these 'clean' technologies advance, the sooner their prices will fall so they will become affordable to developing countries, where more generating capacity will be needed, where the health of the poor is jeopardised by smoky stoves burning wood or dung, and where there would otherwise be pressure to build coal-fired power stations.

The Sun provides five thousand times more energy to the Earth's surface than our total human demand for energy. It shines especially intensely on Asia and Africa where energy demand is predicted to rise fastest. Unlike fossil fuel, it produces no pollution, and no miners get killed. Unlike nuclear fission, it leaves no radioactive waste. Solar energy is already competitive for the thousands of villages in India and Africa that are off the grid. But on a larger scale it remains more expensive than fossil fuels and only becomes economically viable due to subsidies or feed-in tariffs. But eventually these subsidies have to stop.

If the Sun (or wind) is to become the primary source of our energy, there must be some way to store it, so there's still a supply at night and on days when the wind doesn't blow. There's already a big investment in improving batteries and scaling them up. In late 2017 Elon Musk's SolarCity company installed an array of lithium-ion batteries with 100 megawatts capacity at a location in south Australia. Other energy-storage possibilities include thermal storage, capacitors, compressed air, flywheels, molten salt, pumped hydro, and hydrogen.

The transition to electric cars has given an impetus to battery technology (the requirements for car batteries are more demanding than for those in households or 'battery farms', in terms of weight and recharging speed). We'll need high-voltage direct current (HVDC) grids to transmit efficiently over large distances. In the long run these grids should be transcontinental—carrying solar energy from North Africa and Spain to the less sunny northern Europe, and east–west to smooth peak demand over different time zones in North America and Eurasia.

It would be hard to think of a more inspiring challenge for young engineers than devising clean energy systems for the world.

Other methods of power generation apart from the Sun and wind have geographical niches. Geothermal power is readily available in Iceland; wave power may be feasible but is of course as erratic as wind. Harnessing the energy in the tides seems attractive—they rise and fall predictably—but it is actually unpromising, except in a few places where the topography leads to an especially high tidal range. The west coast of Britain, with a tidal range of up to 15 metres, is one such place, and there have been feasibility studies of how turbines could extract energy from the fast tidally induced flows around some capes and promontories. A barrage placed across the wide estuary of the River Severn could yield as much power as several nuclear power stations. But this proposal remains controversial because of concerns about its ecological impact. An alternative scheme involves tidal lagoons, created by building embankments to close off areas of sea several miles across. The difference between the sea level inside and outside is used to drive turbines. These lagoons have the virtue that the capital cost is in low-tech and long-lived earthworks, which could be amortised over centuries.

Current projections suggest that it may be several decades before clean energy sources provide all our needs, especially in the developing world. If, for instance, solar energy and storage by hydrogen and batteries are inadequate (and these seem currently the best bets), then backup will still be needed in midcentury. Gas power would be acceptable if it were combined with carbon sequestration (carbon capture and storage, CCS) whereby the $CO_2$ is extracted from the exhaust gases at the power station and then transported and permanently stored underground.

Some claim that it would be advantageous to actually cut the $CO_2$ concentration back down to its preindustrial level—to sequester not just the future emission from power stations but also to 'suck out' what has been emitted in the past century. The case for this isn't obvious. There's nothing 'optimal' about the world's twentieth-century climate—what's damaging is that the anthropogenic rate of change has been faster than the natural changes in the past, and therefore not easy for us or the natural world to adjust to. But if this reduction were thought worthwhile, there are two ways of achieving it. One is direct extraction from the atmosphere; this is possible, but inefficient as $CO_2$ is only 0.02 percent of the air. Another technique is to grow crops, which of course soak up $CO_2$ from the atmosphere, use them as biofuels, and then capture (and bury) the $CO_2$ that is re-emitted in the power station when they are burned. This is fine in principle but is problematic because of the amount of land needed to grow the fuel (which would otherwise be available for food—or conserved as natural forest), and because the permanent sequestering of the billions of tons of $CO_2$ isn't straightforward. A higher-tech variant would use 'artificial leaves' to incorporate $CO_2$ directly into fuel.

What is the role of nuclear power? I myself would favour the United Kingdom and the United States having at least a replacement generation of power stations. But the hazards of a nuclear accident, even if improbable, cause anxiety; public and political opinion is volatile. After the Fukushima Daiichi disaster in 2011, antinuclear sentiment surged not only (unsurprisingly) in Japan but also in Germany. Moreover, one cannot feel comfortable about a worldwide programme of nuclear power unless internationally regulated fuel banks are established to provide enriched uranium and remove and store the waste—plus a strictly enforced safety code to guard against risks analogous to those from subprime airlines, and a firm nonproliferation agreement to prevent diversion of radioactive material towards weapon production.

Despite the ambivalence about widespread nuclear energy, it's worthwhile to boost R&D into a variety of fourth-generation concepts, which could prove to be more flexible in size, and safer. The industry has been relatively dormant for the last twenty years, and current designs date back to the 1960s or earlier. In particular, it is worth studying the economics of standardised small modular reactors which could be built in substantial numbers and are small enough to be assembled in a factory before being transported to a final location. Moreover, some designs from the 1960s deserve reconsideration—in particular, the thorium-based reactor, which has the advantage that thorium is more abundant in the Earth's crust than uranium, and also produces less hazardous waste.

Attempts to harness nuclear fusion—the process that powers the Sun—have been pursued ever since the 1950s, but the history encompasses receding horizons; commercial fusion power is still at least thirty years away. The challenge is to use magnetic forces to confine gas at a temperature of millions of degrees—as hot as the centre of the Sun—and to devise materials to contain the reactor that can survive prolonged irradiation. Despite its cost, the potential payoff from fusion is so great that it is worth continuing to develop experiments and prototypes. The largest such effort is the International Thermonuclear Experimental Reactor (ITER), in France. Similar projects, but on a smaller scale, are being pursued in Korea, the United Kingdom, and the United States. An alternative concept, whereby converging beams from immense lasers zap and implode tiny deuterium pellets, is being pursued at the Lawrence Livermore National Laboratory in the United States, but this National Ignition Facility is primarily a defence project that will provide lab-scale substitutes for H-bomb tests; the promise of controlled fusion power is a political fig leaf.

A 'dread factor', and a feeling of helplessness, exaggerates public fear of radiation. As a consequence, all fission and fusion projects are impeded by disproportionate concern about even very low radiation levels.

The Japanese tsunami in 2011 claimed thirty thousand lives, mainly through drowning. It also destroyed the Fukushima Daiichi nuclear power stations, which were inadequately protected against a fifteen-metre-high wall of water, and suboptimally designed (for instance, the emergency generators were located low down, and were inactivated by flooding). Consequently, radioactive materials leaked and spread. The surrounding villages were evacuated in an uncoordinated way—initially just those within three kilometres of the power stations, then twenty kilometres, and then thirty—and with inadequate regard for the asymmetric way the wind was spreading the contamination. Some evacuees had to move three times. And some villages remain uninhabited, with devastating consequences for the lives of longtime residents. Indeed, the mental trauma, and other health problems such as diabetes, have proved more debilitating than the radiation risk. Many evacuees, especially the elderly ones, would be prepared to accept a substantially higher cancer risk in return for the freedom to live out their days in familiar surroundings. They should have that option. (Likewise, the mass evacuations after the Chernobyl disaster weren't necessarily in the best interests of those who were displaced.)

Overstringent guidelines about the dangers of low-level radiation worsen the entire economics of nuclear power. After the decommissioning of the Dounreay experimental 'fast breeder' reactor in the north of Scotland, billions of pounds are being spent on an 'interim cleanup' between now and the 2030s, to be followed by further expense spread over several further decades. And nearly 100 billion pounds is budgeted, over the next century, to restore to "greenfields" the Sellafield nuclear installations in England. Another policy concern is this: were a city centre to be attacked by a 'dirty bomb' (a conventional chemical explosion laced with radioactive material), some evacuation would be needed. But, just as in Fukushima, present guidelines would lead to a response that was unduly drastic, both in the extent and the duration of the evacuation. The immediate aftermath of a nuclear incident is not the optimum time for a balanced debate. That is why this topic needs a new assessment now and wide dissemination of clear and appropriate guidelines.

*   *   *

What will actually happen on the climate front? My pessimistic guess is that political efforts to decarbonise energy production won't gain traction, and that the $CO_2$ concentration in the atmosphere will increase at an accelerating rate through the next twenty years, even if the Paris pledges are honoured. But by then we'll know with far more confidence—from a longer time base of data, and from better modelling—just how strong the feedback from water vapour and clouds actually is. If the 'climate sensitivity' is low, we'll relax. But if it's high, and climate consequently seems on an irreversible trajectory into dangerous territory (tracking the steepest of the temperature rise scenarios in the fifth IPCC report), there may then be a pressure for 'panic measures'. This could involve a 'plan B'—being fatalistic about continuing dependence on fossil fuels but combating the effects of releasing $CO_2$ into the atmosphere via a massive investment in carbon capture and storage at all fossil-fuel-powered power stations.

More controversially, the climate could be actively controlled by geoengineering.[20] The 'greenhouse warming' could be counteracted by (for instance) putting reflecting aerosols in the upper atmosphere, or even vast sunshades in space. It seems feasible to throw enough material into the stratosphere to change the world's climate—indeed, what is scary is that this might be within the resources of a single nation, or perhaps even a single corporation. The political problems of such geoengineering may be overwhelming. There could be unintended side effects. Moreover, the warming would return with a vengeance if the countermeasures were ever discontinued, and other consequences of rising $CO_2$ levels (especially the deleterious effects of ocean acidification) would be unchecked.

Geoengineering of this kind would be an utter political nightmare; not all nations would want to adjust the thermostat in the same way. Very elaborate climatic modelling would be needed in order to calculate the regional impacts of any artificial intervention. It would be a bonanza for lawyers if an individual or a nation could be blamed for bad weather! (Note, however, that a different kind of remedy—direct extraction of $CO_2$ from the atmosphere—wouldn't arouse disquiet. This doesn't now seem economically feasible, but it is unobjectionable because it would merely be undoing the geoengineering that humans have already perpetrated through burning fossil fuels.)

Despite its unappealing features, geoengineering is worth exploring to clarify which options make sense and perhaps quench undue optimism about a technical 'quick fix' for our climate. It would also be wise to sort out the complex governance issues raised—and to ensure that these are clarified before climate change becomes so serious that there is pressure for urgent action.

As emphasised in the introduction, this is the first era in which humanity can affect our planet's entire habitat: the climate, the biosphere, and the supply of natural resources. Changes are happening on a timescale of decades. This is far more rapid than the natural changes that occurred throughout the geo-logical past; on the other hand, it is slow enough to give us, collectively or on a national basis, time to plan a response—to mitigate or adapt to a changing climate and modify lifestyles. Such adjustments are possible in principle—though a depressing theme threading though this book is the gap between what is technically desirable and what actually occurs.

We should be evangelists for new technologies—without them we'd lack much of what makes our lives better than the lives of earlier generations. Without technology the world can't provide food, and sus-tainable energy, for an expanding and more demand-ing population. But we need it to be wisely directed. Renewable energy systems, medical advances, and high-tech food production (artificial meat, and so forth) are wise goals; geoengineering techniques probably are not. However, scientific and technical breakthroughs can happen so fast and unpredictably that we may not properly cope with them; it will be a challenge to harness their benefits while avoiding the downsides. The tensions between the promises and the hazards of new technology are the theme of the next chapters.

# 2

# HUMANITY'S FUTURE ON EARTH

## 2.1. BIOTECH

Robert Boyle is best remembered today for 'Boyle's law', relating the pressure and density of gases. He was one of the 'ingenious and curious gentlemen' who, in 1660, founded the Royal Society of London, which still exists as the United Kingdom's academy of sciences. These men (and there were no women among them) would have called themselves 'natural philosophers' (the term 'scientist' didn't exist until the nineteenth century); in the words of Francis Bacon, whose writings deeply influenced them, they were 'merchants of light', pursuing enlightenment for its own sake; but they were also practical men engaged in the problems of their time, and aiming (to quote Bacon again) at 'the relief of Man's estate'.

Boyle was a polymath. After he died in 1691, a handwritten note was found among his papers, with a 'wish list' of inventions that would benefit humankind.[1] In the quaint idiom of his time, he envisaged some advances that have now been achieved, and some that still elude us more than three centuries later. Here is part of his list:

The Prolongation of Life.

The Recovery of Youth, or at least some of the Marks of it, as new Teeth, new Hair colour'd as in youth

The art of flying

The Art of Continuing long under water, and exercising functions freely there

Great Strength and Agility of Body exemplify'd by that of Frantick Epileptick and Hystericall persons

The Acceleration of the Production of things out of Seed

The making of Parabolicall and Hyperbolicall Glasses

The practicable and certain way of finding Longitudes

Potent Druggs to alter or Exalt Imagination,
  Waking, Memory, and other functions,
  and appease pain, procure innocent sleep,
  harmless dreams, etc
A perpetuall Light
The Transmutation of Species in Mineralls,
  Animals, and Vegetables
The Attaining Gigantick Dimensions
Freedom from Necessity of much Sleeping
  exemplify'd by the Operations of Tea and
  what happens in Mad-Men and stimulants to
  keep you awake.[2]

Anyone from Boyle's seventeenth century would be astonished by the modern world—far more than a Roman would have been by Boyle's world. Moreover, many changes are still accelerating. Novel technologies—bio, cyber, and AI—will be transformative in ways that are hard to predict even a decade ahead. These technologies may offer new solutions to the crises that threaten our crowded world; on the other hand, they may create vulnerabilities that give us a bumpier ride through the century. Further progress will depend on findings that come fresh from research laboratories, so the speed of advance is especially unpredictable—a contrast with, for instance, nuclear power, which is based on twentieth-century physics, and with the nineteenth-century transformations wrought by steam and electricity.

A 'headline' trend in biotech has been the plummeting cost of sequencing the genome. The 'first draft of the human genome' was 'big science'—an international project with a three-billion-dollar budget. Its completion was announced at a press conference at the White House in June 2000. But the cost has by 2018 fallen below one thousand dollars. Soon it will be routine for all of us to have our genome sequenced—raising the question of whether we really want to know if we carry the genes that give us a propensity to particular diseases.[3]

But now there's a parallel development—the faster and cheaper ability to *synthesise* genomes. Already in 2004, the polio virus was synthesised—a portent of things to come. In 2018, the technique is now far advanced. Indeed, Craig Venter, the American biotechnologist and businessman, is developing a gene synthesiser that is, in effect, a 3D printer for genetic codes. Even if it is only able to reproduce short genomes, this could have varied applications. The 'code' for a vaccine could be electronically transmitted around the world—allowing instant global distribution of a vaccine created to counter a new epidemic.

People are typically uneasy about innovations that seem 'against nature' and that pose risks. Vaccination and heart transplants, for instance, aroused controversy in the past. More recently, concern has focused on embryo research, mitochondrial transplants, and stem cells. I followed closely the debate in the United Kingdom that led to legislation allowing experiments on embryos up to fourteen days old. This debate was well handled; it was characterised by constructive engagement between the researchers, the parliamentarians, and the wider public. There was opposition from the Catholic Church, some of whose representatives circulated pamphlets depicting a fourteen-day-old embryo as a structured 'homunculus'. Scientists rightly emphasised how misleading this was; such an early-stage embryo is actually a microscopic and still undifferentiated group of cells. But the more sophisticated opponents would respond, 'I know that, but it's still sacred'—and to that belief science can offer no counterargument.

In contrast, the debate on genetically modified (GM) crops and animals was handled less well in the United Kingdom. Even before the public was fully engaged there was a standoff between Monsanto, a giant agrochemical corporation, and environmentalists. Monsanto was accused of exploiting farmers in the developing world by requiring them to purchase seeds annually. The wider public was influenced by a newspaper campaign against 'Frankenstein Foods'. There was a 'yuck' factor when we learned that scientists could 'create' rabbits that glow in the dark—an aggravated version of the distaste many of us feel at the exploitation of circus animals. Despite the fact that GM crops have been consumed by three hundred million Americans for an entire decade without manifest damage, they are still severely restricted within the European Union. And, as mentioned in section 1.3, providing GM foodstuffs to undernourished children to combat dietary deficiencies has been impeded by anti-GM campaigners. But there are genuine concerns that reduced genetic diversity in crucial crops (wheat, maize, and such) might render the world's food supply more vulnerable to plant diseases.

The new gene-editing technology, CRISPR/Cas9, could modify gene sequences in a more acceptable manner than earlier techniques. CRISPR/Cas9 makes small changes in the sequences of DNA to suppress (or alter the expression of) damaging genes. But it doesn't 'cross the species barrier'. In humans, this most benign and uncontroversial use of gene editing removes single genes that cause specific disease.

In vitro fertilisation (IVF) already provides a less invasive way than CRISPR/Cas9 to weed out damaging genes. In this procedure, after hormone treatment to induce ovulation, several eggs are harvested, fertilised in vitro, and allowed to develop to an early stage. A cell from each embryo is then tested for the presence of the undesirable gene, and one that is free of it is then implanted for a normal pregnancy.

A different technique is now available that can replace a particular category of faulty genes. Some of the genetic material in a cell is found in organelles called mitochondria; these are separate from the cell's nucleus. If a faulty gene is mitochondrial, it is possible to replace it with mitochondria from another female—leading to 'three-parent babies'. This technique was legalised by the United Kingdom's parliament in 2015. The next step would be to use gene editing on DNA in the cell nucleus.

In the public's mind, a sharp distinction exists between artificial medical interventions that remove something harmful and deploying similar techniques to offer 'enhancement'. Most characteristics (size, intelligence, and such) are determined by an aggregate of many genes. Only when the DNA of millions of people is available will it become possible (using pattern-recognition systems aided by AI) to identify the relevant combination of genes. In the short term, this knowledge could be used to inform embryo selection for IVF. But modification or redesign of the genome is a more remote (and of course more risky and dubious) prospect. Not until this can be done—and until DNA with the required prescription can be artificially sequenced— will 'designer babies' become conceivable (in both senses of that word!). Interestingly, it is unclear how much parental desire there would be for offspring 'enhanced' in this fashion (as opposed to the more feasible single-gene editing needed to remove propensities towards specific diseases or disabilities). In the 1980s, the Repository for Germinal Choice was established in California with the aim of enabling the conception of 'designer babies'; it was a sperm bank, dubbed the Nobel prize sperm bank, with only 'elite' donors, including William Shockley, co-inventor of the transistor and a Nobel prize winner, who achieved notoriety later in life for being a proponent of eugenics. He was surprised—though most of us were probably gratified—that there was no great demand.

The advances in medicine and surgery already achieved—and those we can confidently expect in the coming decades—will be acclaimed as a net blessing. They will nonetheless sharpen up some ethical issues—in particular, they will render more acute the dilemmas involved in treating those at the very beginning and the end of their lives. An extension of our healthy lifespan will be welcome. But what is becoming more problematic is the growing gap between how long we will survive in healthy old age and how much longer some kind of life can be extended by extreme measures. Many of us would choose to request non-resuscitation, and solely palliative treatment, as soon as our quality of life and our prognosis dropped below a threshold. We dread clinging on for years in the grip of advanced dementia—a drain on resources, and on the sympathy of others. Similarly, one must question whether the efforts to save extremely premature or irreversibly damaged babies has gone too far. In late 2017, for instance, a team of UK surgeons tried—with immense commitment and dedication—to save an infant born with her heart outside her body.

Belgium, Holland, Switzerland, and several US states have legalised 'assisted dying'—thereby ensuring that a person of sound mind with a terminal disease can be helped to a peaceful death. Relatives, or physicians and their helpers, can carry out the necessary procedures without being threatened with criminal prosecution for 'aiding a suicide'. Nothing similar yet has parliamentary approval in the United Kingdom. The objections are based on fundamental religious grounds, on the view that participation in such acts is contrary to a doctor's ethical code, and on worries that vulnerable people might feel pressured to take this course by their families or by undue concern about placing burdensome demands on others. This inaction in the United Kingdom persists, despite 80 percent public support for 'assisted dying'. I am firmly in that 80 percent. Knowledge that this option was available would comfort many more than the number who would actually make use of it. Modern medicine and surgery obviously benefit most of us, for most of our lives, and we can expect further advances in the coming decades that can prolong healthy lives. Nonetheless, I expect (and hope) there will be enhanced pressure for legalising euthanasia under regulated conditions.

Another consequence of medical advances is the blurring of the transition between life and death. Death is now normally defined as 'brain death'—the stage when all measurable signs of brain activity become extinguished. This is the criterion transplant surgeons use in deciding when they can properly 'harvest' a body's organs. But the line is being blurred further by proposals that the heart can be artificially restarted after 'brain death', simply to keep the targeted organs 'fresh' for longer. This introduces further moral ambiguity into transplant surgery. Already, 'agents' are inducing impoverished Bangladeshis to sell a kidney or other organ that will be resold with a huge markup to benefit wealthy potential recipients. And we've all seen distressing TV footage of a mother with a sick child pleading that she is 'desperate for a donor'—desperate, in other words, for another child to die, perhaps from a fatal accident, to supply the needed organ. These moral ambiguities, together with a shortage of organ donors, will continue (and indeed be aggravated) until xenotransplantation—harvesting organs for human use from pigs or other animals—becomes routine and safe. Better still (though more futuristically), techniques akin to those being developed in order to make artificial meat could enable 3D printing of replacement organs. These are advances that should be prioritised.

Advances in microbiology—diagnostics, vaccines, and antibiotics—offer prospects of sustaining health, controlling disease, and containing pandemics. But these benefits have triggered a dangerous 'fight back' by the pathogens themselves. There are concerns about antibiotic resistance whereby bacteria evolve (via speeded-up Darwinian selection) to be immune against the antibiotics used to suppress them. This has led, for example, to a resurgence in tuberculosis (TB). Unless new antibiotics are developed, the risks of (for instance) untreatable postoperative infections will surge to the level of a century ago. In the short term, it's urgent to prevent the overuse of antibiotics—for instance, in cattle in the United States—and to incentivise the development of new antibiotics, even though these are less profitable to pharmaceutical companies than the drugs that control long-term conditions.

And studies of viruses, carried out in the hope of thereby developing improved vaccines, have controversial aspects. For instance, in 2011, two research groups, one in Holland and another in Wisconsin, showed that it was surprisingly easy to make the H5N1 influenza virus both more virulent and more transmissible—in contrast to the natural trend for these two features to be anticorrelated. The justification adduced for these experiments was that by staying one step ahead of natural mutations, it would be easier to prepare vaccines in good time. But, to many, this benefit was outweighed by the enhanced risks of unintentional release of dangerous viruses, plus the wider dissemination of techniques that could be helpful to bioterrorists. In 2014, the US government ceased funding these so-called gain of function experiments—but in 2017 this ban was relaxed. In 2018 a paper was published reporting the synthesis of the horsepox virus—with the implication that a smallpox virus could be similarly synthesised.[4] Some questioned the justification for this research, carried out by a group in Edmonton, Alberta, because a safe smallpox virus already exists and is stockpiled; others argued that even if the research were justifiable, publication was a mistake.

As already mentioned, experiments using CRISPR/Cas9 techniques on human embryos raise ethical concerns. And the rapid advance of biotech will bring up further instances where there's concern about the safety of experiments, the dissemination of 'dangerous knowledge', and the ethics of how it's applied. Procedures that affect not just an individual but his or her progeny—altering the germ line—are disquieting. There has, for instance, been an attempt, with 90 percent success, to make sterile, and thereby wipe out, the species of mosquito that spreads the Dengue and Zika strains of virus. In the United Kingdom, a 'gene drive' has been invoked to remove grey squirrels—regarded as a 'pest' that threatens the cuddlier red variety. (A more benign tactic is to engineer the red squirrel so that it can resist the parapoxvirus that is spread by the grey squirrels.) Similar techniques are being proposed that could preserve the unique ecology of the Galapagos Islands by eliminating invasive species—black rats in particular. But it's worth noting that in a recent book, *Inheritors of the Earth*, Chris Thomas, a distinguished ecologist, argues that the spread of species can often have a positive impact in ensuring a more varied and robust ecology.[5]

In 1975, in the early days of recombinant DNA research, a group of leading molecular biologists met at the Asilomar Conference Grounds in Pacific

Grove, California, and agreed on guidelines defining what experiments should and should not be done. This seemingly encouraging precedent has triggered several meetings, convened by national academies, to discuss recent developments in the same spirit. But today, more than forty years after the first Asilomar meeting, the research community is far more broadly international, and more influenced by commercial pressures. I'd worry that whatever regulations are imposed, on prudential or ethical grounds, cannot be enforced worldwide—any more than the drug laws can, or tax laws. Whatever can be done will be done by someone, somewhere. And that's a nightmare. In contrast to the elaborate and conspicuous special-purpose equipment needed to create a nuclear weapon, biotech involves small-scale dual-use equipment. Indeed, biohacking is burgeoning even as a hobby and competitive game.

Back in 2003 I was worried about these hazards and rated the chance of bio error or bio terror leading to a million deaths as 50 percent by 2020. I was surprised at how many of my colleagues thought a catastrophe was even more likely than I did. More recently, however, psychologist/author Steven Pinker took me up on that bet, with a two-hundred-dollar stake. This is a bet that I fervently hope to lose, but I was not surprised that the author of *The Better Angels of Our Nature*[6] should take an optimistic line. Pinker's fascinating book is infused with optimism. He quotes statistics pointing to a gratifying downward trend in violence and conflict—a decline that has been obscured by the fact that global news networks report disasters that would have been unreported in earlier times. But this trend can lull us into undue confidence. In the financial world, gains and losses are asymmetric; many years of gradual gains can be wiped out by a sudden loss. In biotech and pandemics, the risk is dominated by the rare but extreme events. Moreover, as science empowers us more, and because our world is so interconnected, the magnitude of the worst potential catastrophes has grown unprecedentedly large, and too many are in denial about them.

By the way, the societal fallout from pandemics would be far higher than in earlier centuries. European villages in the mid-fourteenth century continued to function even when the Black Death almost halved their populations; the survivors were fatalistic about a massive death toll. In contrast, the feeling of entitlement is so strong in today's wealthier countries that there would be a breakdown in the social order as soon as hospitals overflowed, key workers stayed at home, and health services were overwhelmed. This could occur when those infected were still a fraction of 1 percent. The fatality rate would, however, probably be highest in the megacities of the developing world.

Pandemics are an ever-present natural threat, but is it just scaremongering to raise concerns about human-induced risks from bio error or bio terror? Sadly, I don't think it is. We know all too well that technical expertise doesn't guarantee balanced rationality. The global village will have its village idiots and they'll have global range. The spread of an artificially released pathogen can't be predicted or controlled; this realisation inhibits the use of bioweapons by governments, or even by terrorist groups with specific well-defined aims (which is why I focused on nuclear and cyber threats in section 1.2). So, my worst nightmare would be an unbalanced 'loner', with biotech expertise, who believed, for instance, that there were too many humans on the planet and didn't care who, or how many, were infected. The rising empowerment of tech-savvy groups (or even individuals) by bio- as well as cybertechnology will pose an intractable challenge to governments and aggravate the tension among freedom, privacy, and security. Most likely there will be a societal shift towards more intrusion and less privacy. (Indeed, the rash abandon with which people put their intimate details on Facebook, and our acquiescence in ubiquitous CCTV [video surveillance], suggests that such a shift would meet surprisingly little resistance.)

Bio error and bio terror are possible in the near term—within ten or fifteen years. And in the longer term they will be aggravated as it becomes possible to 'design' and synthesise viruses—the 'ultimate' weapon would combine high lethality with the transmissibility of the common cold.

What advances might biologists bring us in 2050 and beyond? Freeman Dyson projects a time when children will design and create new organisms just as routinely as his generation played with chemistry sets.[7] If one day it became possible to 'play God on a kitchen table', our ecology (and even our species) might not survive long unscathed. Dyson, however, isn't a biologist; he's one of the twentieth century's leading theoretical physicists. But unlike many such people he's a divergent and speculative thinker— often expressing a contrarian bent. For instance, back in the 1950s he was part of a group that explored the speculative concept 'Project Orion'. The group's aim was to achieve interstellar travel with spaceships powered by exploding H-bombs (nuclear pulse propulsion) behind the (well shielded) vehicle. Even in 2018, Dyson remains sceptical about the need for an urgent response to climate change.

Research on aging is being seriously prioritised. Will the benefits be incremental? Or is aging a 'disease' that can be cured? Serious research focuses on telomeres, stretches of DNA at the ends of chromosomes, which shorten as people age. It's been possible to achieve a tenfold increase in the lifespan of nematode worms, but the effect on more complex animals is less dramatic. The only effective way to extend the life of rats is by giving them a near-starvation diet. But there's one unprepossessing creature, the naked mole rat, which may have some special biological lessons for us; some live for more than thirty years—several times longer than the lifespan of other small mammals.

Any major breakthrough in life extension for humans would alter population projections in a drastic way; the social effects, obviously huge, would depend on whether the years of senility were prolonged too, and on whether the age of women at menopause would increase with total lifespan. But various types of human enhancement via hormonal treatment may become possible as the human endocrine system is better understood—and some degree of life extension is likely to be among these enhancements. As with so much technology, priorities are unduly slanted towards the wealthy. And the desire for a longer lifespan is so powerful that it creates a ready market for exotic therapies with untested efficacy. Ambrosia, a 2016 start-up, offers Silicon Valley executives a transfusion of 'young blood'. Another recent craze was metformin, a drug intended to treat diabetes, but which is claimed to stave off dementia and cancer; others extol the benefits of placental cells. Craig Venter has a company called Human Longevity, which received $300 million in start-up funds. This goes beyond 23andMe (the firm that analyses our genome well enough to reveal interesting results about our vulnerability to some diseases, and about our ancestry). Venter aims to analyse the genomes of the thousands of species of 'bugs' in our gut. It is believed (very plausibly) that this internal 'ecosystem' is crucial to our health.

The 'push' from Silicon Valley towards achieving 'eternal youth' stems not only from the immense surplus wealth that's been accumulated there, but also because it's a place with a youth-based culture. Those older than thirty are thought to be 'over the hill'. The futurist Ray Kurzweil speaks zealously of attaining a metaphorical 'escape velocity'—when medicine advances so fast that life expectancy rises by more than a year in each year, offering potential immortality. He ingests more than one hundred supplements a day—some routine, some more exotic. But he's worried that 'escape velocity' may not be achieved within his 'natural' lifetime. So, he wants his body frozen until this nirvana is reached.

I was once interviewed by a group of 'cryonics' enthusiasts—based in California—called the 'society for the abolition of involuntary death'. I told them I'd rather end my days in an English churchyard than a Californian refrigerator. They derided me as a 'deathist'—really old fashioned. I was surprised to learn later that three academics in England (though I'm glad to say not from my university) had signed up for 'cryonics'. Two have paid the full whack; the third has taken the cut-price option of contracting for just his head to be frozen. The contract is with a company called Alcor in Scottsdale, Arizona. These colleagues are realistic enough to accept that the chance of resurrection may be small, but they point out that without this investment the chance is zero. So they wear a medallion displaying instructions to immediately freeze them when they die and replace their blood with liquid nitrogen.

It is hard for most of us mortals to take this aspiration seriously; moreover, if cryonics had a real prospect of success, I don't think it would be admirable either. If Alcor didn't go bust and dutifully maintained the refrigeration and stewardship for the requisite centuries, the corpses would be revived into a world where they would be strangers—refugees from the past. Perhaps they would be indulgently treated, as we feel we should treat (for instance) distressed asylum seekers, or Amazonian tribal people who've been forced from their familiar habitat. But the difference is that the 'thawed-out corpses' would be burdening future generations by choice; so, it's not clear how much consideration they would deserve. This is reminiscent of a similar dilemma that may not always be science fiction, even if it should remain so: cloning a Neanderthal. One of the experts (a Stanford professor) queried: 'Would we put him in a zoo or send him to Harvard?'

## 2.2. CYBERTECHNOLOGY, ROBOTICS, AND AI

Cells, viruses, and other biological microstructures are essentially 'machines' with components on the molecular scale—proteins, ribosomes, and so forth. We owe the dramatic advances in computers to the fast-advancing ability to manufacture electronic components on the nanoscale, thereby allowing almost biological-level complexity to be packed into the processors that power smartphones, robots, and computer networks.

Thanks to these transformative advances, the internet and its ancillaries have created the most rapid 'penetration' of new technology in history—and also the most fully global. Their spread in Africa and China proceeded faster than nearly all 'expert' predictions. Our lives have been enriched by consumer electronics and web-based services that are affordable by literally billions. And the impact on the developing world is emblematic of how optimally applied science can transform impoverished regions. Broadband internet, soon to achieve worldwide reach via low-orbiting satellites, high-altitude balloons, or solar-powered drones, should further stimulate education and the adoption of modern health care, agricultural methods, and technology; even the poorest can thereby leapfrog into a connected economy and enjoy social media— even though many are still denied the benefits of nineteenth-century technological advances such as proper sanitation. People in Africa can use smart-phones to access market information, make mobile payments, and so forth; China has the most auto-mated financial system in the world. These developments have a 'consumer surplus' and generate enterprise and optimism in the developing world. And such benefits have been augmented by effective programmes aiming to eliminate infectious diseases such as malaria. According to the Pew Research Center, 82 percent of Chinese people and 76 percent of Indians believe that their children will be better off than they themselves are today.

Indians now have an electronic identity card that makes it easier for them to register for wel-fare benefits. This card doesn't need passwords. The vein pattern in our eyes allows the use of 'iris recognition' software—a substantial improvement on fingerprints or facial recognition. This is pre-cise enough to unambiguously identify individuals, among the 1.3 billion Indians. And it is a foretaste of the benefits that can come from future advances in AI.

Speech recognition, face recognition, and similar applications use a technique called generalised machine learning. This operates in a fashion that resembles how humans use their eyes. The 'visual' part of human brains integrates information from the retina through a multistage process. Successive layers of processing identify horizontal and vertical lines, sharp edges, and so forth; each layer processes information from a 'lower' layer and then passes its output to other layers.[8]

The basic machine-learning concepts date from the 1980s; an important pioneer was the Anglo-Canadian Geoff Hinton. But the applications only really 'took off' two decades later, when the steady operation of Moore's law—a doubling of computer speeds every two years—led to machines with a thousand times faster processing speed. Computers use 'brute force' methods. They learn to translate by reading millions of pages of (for example) multilingual European Union documents (they never get bored!). They learn to identify dogs, cats, and human faces by 'crunching' through millions of images viewed from different perspectives.

Exciting advances have been spearheaded by DeepMind, a London company now owned by Google. DeepMind's cofounder and CEO, Demis Hassabis, has had a precocious career. At thirteen he was ranked the number two chess champion in the world for his category. He qualified for admission to Cambridge at fifteen but delayed admission for two years, during which time he worked on computer games, including conceiving the highly successful Theme Park. After studying computer science at Cambridge, he started a computer games company. He then returned to academia and earned a PhD at University College London, followed by postdoctoral work on cognitive neuroscience. He studied the nature of episodic memory and how to simulate groups of human brain cells in neural net machines.

In 2016, DeepMind achieved a remarkable feat—its computer beat the world champion of the game of Go. This may not seem a 'big deal' because it's been more than twenty years since IBM's supercomputer Deep Blue beat Garry Kasparov, the world chess champion. But it was a 'game change' in the colloquial as well as literal sense. Deep Blue had been programmed by expert players. In contrast, the AlphaGo machine gained expertise by absorbing huge numbers of games and playing itself. Its designers don't know how the machine makes its decisions. And in 2017 AlphaGo Zero went a step further; it was just given the rules—no actual games—and learned completely from scratch, becoming world-class within a day. This is astonishing. The scientific paper describing the feat concluded with the thought that

> humankind has accumulated Go knowledge from millions of games played over thousands of years, collectively distilled into patterns, proverbs and books. In the space of a few days, starting tabula rasa, AlphaGo Zero was able to rediscover much of this Go knowledge, as well as novel strategies that provide new insight into the oldest of games.[9]

Using similar techniques, the machine reached Kasparov-level chess competence within a few hours, without expert input, and similar prowess in the Japanese game of Shogi. A computer at Carnegie Mellon University has learned to bluff and calculate as well as the best professional poker players. But Kasparov himself has emphasised that in games like chess humans offer distinctive 'added value' and that a person plus a machine, in combination, can surpass what either could accomplish separately.

AI earns its advantage over humans through its ability to analyse vast volumes of data and rapidly manipulate and respond to complex input. It excels in optimising elaborate networks, like the electricity grid or city traffic. When the energy management of its large data farms was handed over to a machine, Google claimed energy savings of 40 percent. But there are still limitations. The hardware underlying AlphaGo used hundreds of kilowatts of power. In contrast, the brain of Lee Sedol, AlphaGo's Korean challenger, consumes about thirty watts (like a lightbulb) and can do many other things apart from play board games.

Sensor technology, speech recognition, information searches, and so forth are advancing apace. So (albeit with a more substantial lag) is physical dexterity. Robots are still clumsier than a child in moving pieces on a real chessboard, tying shoelaces, or cutting toenails. But here too there is progress. In 2017, Boston Dynamics demonstrated a fearsome-looking robot called Handel (a successor to the earlier four-legged Big Dog), with wheels as well as two legs, that is agile enough to perform back flips. But it will be a long time before machines outclass human gymnasts—or indeed interact with the real world with the agility of monkeys and squirrels that jump from tree to tree—still less achieve the overall versatility of humans.

Machine learning, enabled by the ever-increasing number-crunching power of computers, is a potentially stupendous breakthrough. It allows machines to gain expertise—not just in game playing, but in recognising faces, translating between languages, managing networks, and so forth—without being programmed in detail. But the implications for human society are ambivalent. There is no 'operator' who knows exactly how the machine reaches a decision. If there is a 'bug' in the software of an AI system, it is currently not always possible to track it down; this is likely to create public concern if the system's 'decisions' have potentially grave consequences for individuals. If we are sentenced to a term in prison, recommended for surgery, or even given a poor credit rating, we would expect the reasons to be accessible to us—and contestable by us. If such decisions were entirely delegated to an algorithm, we would be entitled to feel uneasy, even if presented with compelling evidence that, on average, the machines make better decisions than the humans they have usurped.

Integration of these AI systems has an impact on everyday life—and will become more intrusive and pervasive. Records of all our movements, our interactions with others, our health, and our financial transactions, will be in the 'cloud', managed by a multinational quasi-monopoly. The data may be used for benign reasons (for instance, for medical research, or to warn us of incipient health risks), but its availability to internet companies is already shifting the balance of power from governments to the commercial world. Indeed, employers can now monitor individual workers far more intrusively than the most autocratic or 'control freak' traditional bosses. There will be other privacy concerns. Are you happy if a random stranger sitting near you in a restaurant or on public transportation can, via facial recognition, identify you, and invade your privacy? Or if 'fake' videos of you become so convincing that visual evidence can no longer be trusted?

## 2.3. WHAT ABOUT OUR JOBS?

The pattern of our lives—the way we access information and entertainment, and our social networks—has already changed to a degree that we would hardly have envisioned twenty years ago. Moreover, AI is just at the 'baby stage' compared to what its proponents expect in coming decades. There will plainly be drastic shifts in the nature of work, which not only provides our income but also helps give meaning to our lives and our communities. So, the prime social and economic question is this: Will this 'new machine age' be like earlier disruptive technologies—the railways, or electrification, for instance—and create as many jobs as it destroys? Or is it really different this time?

During the last decade the real wages of unskilled people in Europe and North America fell. So did the security of such people's employment. Despite that, one countervailing factor has offered all of us greater subjective well-being: the consumer surplus offered by the ever more pervasive digital world. Smartphones and laptops have improved vastly. I value internet access far more than I value owning a car, and it's far cheaper.

Clearly, machines will take over much of the work of manufacturing and retail distribution. They can replace many white-collar jobs: routine legal work (such as conveyancing), accountancy, computer coding, medical diagnostics, and even surgery. Many 'professionals' will find their hard-earned skills in less demand. In contrast, some skilled service-sector jobs—plumbing and gardening, for instance—require nonroutine interactions with the external world and so will be among the hardest jobs to automate. To take a much-cited example, how vulnerable are the jobs of three million truck drivers in the United States?

Self-driving vehicles may be quickly accepted in limited areas where they will have the roads to themselves—in designated parts of city centres, or maybe in special lanes on motorways. And there is a potential for using driverless machines in farming and harvesting, operating off road. But what is not so clear is whether automated vehicles will ever be able to operate safely when confronted with all the complexities of routine driving—navigating small, winding roads and sharing city streets with human-driven vehicles and cycles and pedestrians. I think there will be public resistance to this.

Would a fully autonomous car be safer than a car with a human driver? If an object obstructs the road ahead, could it distinguish between a paper bag, a dog, or a child? The claim is that it cannot infallibly do so but will do better than the average human driver. Is that true? Some would say yes. If the cars are wirelessly connected to one another, they would learn faster by sharing experiences.

On the other hand, we should not forget that every innovation is initially risky—think of the early days of railways, or the pioneering use of surgical operations that are now routine. Regarding road safety, here are some figures from the United Kingdom. In 1930, when there were only a million vehicles on the roads, there were more than 7,000 fatalities; in 2017 there were about 1,700 fatalities—a drop by a factor of four, even though there are about thirty times more vehicles on the roads than there were in 1930.[10] The trend is due partly to better roads, but largely to safer cars and, in recent years, to satellite-based navigation systems (satnavs) and other electronic gadgetry. This trend will continue, making driving safer and easier. But fully automatic vehicles sharing ordinary roads with mixed traffic would be a truly disjunctive change. We are justified in being sceptical about how feasible and acceptable this transition would be.

It may be a long time before truck and car drivers are redundant. As a parallel, consider what is happening in civil aviation. Although air travel was once dangerous, it is now amazingly safe. During 2017 there was not a single fatality, worldwide, on any scheduled airliner. Most flying is done on autopilot; a real pilot is needed only in emergencies. But the concern is that he or she may not be alert at the crucial time. The 2009 crash of an Air France plane, en route from Rio de Janeiro, Brazil, to Paris, in the South Atlantic demonstrates what can go wrong: the pilots took too long to resume control when there was an emergency and mistakenly aggravated the problem. On the other hand, suicidal pilots have actually caused devastating crashes that the autopilot couldn't prevent. Will the public ever be relaxed about boarding a plane with no pilot? I doubt it. But pilotless planes may be acceptable for air freight. Small delivery drones have a promising future; indeed, in Singapore, there are plans to replace delivery vehicles at ground level with drones flying above the streets. But even for these, we are too complacent about the risk of collisions, especially if they proliferate. For ordinary cars, software errors and cyberattacks cannot be ruled out. We are already seeing the hackability of the ever more sophisticated software and security systems found in automobiles. Can we confidently protect brakes and steering against hacking?

An oft-quoted benefit of driverless cars is that they will be hired and shared rather than owned. This could reduce the amount of parking space needed in cities—blurring the line between public and private transport. But what is not clear is how far this will go—whether the wish to possess one's own car will indeed disappear. If driverless cars catch on, they will boost road travel at the expense of traditional rail travel. Many people in Europe prefer taking the train for a 200-mile journey; it is less stressful than driving and opens up time to work or read. But if we had an 'electronic chauffeur' who could be trusted for the entire journey, many of us would prefer to travel by car and get door-to-door service. This would reduce demand for long-distance train routes—but at the same time provide an incentive for inventing novel forms of transport, such as intercity hyperloops. Best of all, of course, would be high-grade telecommunications that obviate the need for most nonleisure travel.

The digital revolution generates enormous wealth for an elite group of innovators and for global companies, but preserving a healthy society will require redistribution of that wealth. There is talk of using it to provide a universal income. The snags to implementing this are well known, and the societal disadvantages are intimidating. It would be far better to subsidise the types of jobs for which there is currently a large unmet demand and for which pay and status is unjustly low.

It's instructive to observe (sometimes with bemusement) the spending choices made by those who are not financially constrained. Rich people value personal service; they employ personal trainers, nannies, and butlers. When they're elderly, they employ human caregivers. The criterion for a progressive government should be to provide for everyone the kind of support preferred by the best-off—the ones who now have the freest choice. To create a humane society, governments will need to vastly enhance the number and status of those who carry out caregiving roles; there are currently far too few, and even in wealthy countries caregivers are poorly paid and insecure in their positions. (It's true that robots can take over some aspects of routine care—indeed, we may find it less embarrassing for basic washing, feeding, and bedpan routines to be handled by an automaton. But those who can afford it want the attention of real human beings as well.) And there are other jobs that would make our lives better and could provide worthwhile employment for far more people—for example, gardeners in public parks, custodians, and so forth.

It's not just the very young and very old who need human support. When so much business, including interaction with government, is done via the internet, we should worry about, for instance, a disabled person living alone who needs to access websites online to claim their rightful government benefits, or to order basic provisions. Think of the anxiety and frustration when something goes wrong. Such people will have peace of mind only if there are computer-savvy caregivers to help the bewildered cope with IT, to ensure that they can get help and are not disadvantaged. Otherwise, the 'digitally deprived' will become a new 'underclass'.

It is better when we can all perform socially useful work rather than receive a handout. However, the typical working week could be shortened—to shorter even than France's current thirty-five hours. Those for whom work is intrinsically satisfying are atypical and especially lucky. Most people would welcome shorter hours, which would release more time for entertainment, socialising, and for participation in collective rituals—whether religious, cultural, or sporting.

There will also be a resurgence of arts and crafts. We've seen the emergence of 'celebrity chefs'—even celebrity hairdressers. We'll see more scope for other crafts, and more respect accorded to their most talented exponents. Again, the wealthy, those who have the most freedom of choice, spend heavily on patronising labour-intensive activities.

The erosion of routine work and lifetime careers will stimulate 'life-long learning'. Formal education, based on teaching done in classrooms and lecture halls, is perhaps the most sclerotic sector of societies worldwide. Distance learning via online courses may never replace the experience of attending a residential college that offers personal mentoring and tuition, but it will become a cost effective and more flexible replacement for the typical 'mass university'. There is boundless potential for the model pioneered by the United Kingdom's Open University, a model that is now being spread widely via US organisations like Coursera and edX, where leading academics provide content for online courses. Teachers who do this best can become global online stars. These courses will be enhanced by the personalisation that

AI will increasingly be able to provide. Those who become scientists often attribute their initial motivation to the web or media rather than to classroom instruction.

The lifestyle a more automated world offers seems benign—indeed enticing—and could in principle promote Scandinavian-level satisfaction throughout Europe and North America. However, citizens of these privileged nations are becoming far less isolated from the disadvantaged parts of the world. Unless inequality between countries is reduced, embitterment and instability will become more acute because the poor, worldwide, are now, via IT and the media, far more aware of what they're missing. Technical advances could amplify international disruption. Moreover, if robotics renders it economically viable for wealthy countries to shore up manufacturing within their own borders, the transient but crucial developmental boost that the 'tigers' in the Far East received by undercutting Western labour costs will be denied to the still-poor nations in Africa and the Middle East, rendering the inequalities more persistent.

Also, the nature of migration has changed. A hundred years ago a European or Asian individual's decision to move to North America or Australia required severing ties with his or her indigenous culture and extended family. There was therefore an incentive to integrate into the new society. In contrast, daily video calls and social media contacts now enable immigrants, if they so choose, to remain embedded in the culture of their homeland, and affordable intercontinental travel can sustain personal contacts.

National and religious loyalties and divisions will persist (or even be strengthened by internet echo chambers) despite greater mobility and less sentimentality about 'place'. Nomads of the technocratic world will expand in numbers. The impoverished will see 'following the money' as their best hope— migrating legally or illegally. International tensions will get more acute.

If there is indeed a growing risk of conflicts triggered by ideology or perceived unjust inequality, it will be aggravated by the impact of new technology on warfare and terrorism. For the last decade at least, we've seen TV reports of drones or rockets attacking targets in the Middle East. They are controlled from bunkers in the continental United States—by individuals even more remote from the consequences of their actions than aircrews carrying out bombing raids. The ethical queasiness this engenders is somewhat allayed by claims that higher-precision targeting reduces collateral damage. But at least there is a human 'in the loop' who decides when and what to attack. In contrast, there is now the possibility of autonomous weapons, which can seek out a target—using facial recognition to identify individuals and then kill them. This would be a precursor to automated warfare—a development that raises deep concerns. Near-term possibilities include automated machine guns; drones; and armoured vehicles or submarines that can identify targets, decide whether to open fire, and learn as they go.

There is rising concern about 'killer robots'. In August 2017, the heads of one hundred leading companies in this field signed an open letter calling on the United Nations to outlaw 'lethal autonomous weapons', just as international conventions constrain the use of chemical and biological weapons.[11] The signatories warn about an electronic battlefield 'at a scale greater than ever, and at timescales faster than humans can comprehend'. How effective any such treaty would be remains unclear; just as in the case of bioweapons, nations may pursue these technologies for allegedly 'defensive' motives, and through fear that rogue nations or extremist groups would go ahead with such developments anyway.

These are near-term concerns, for which the key technologies are already understood. But let's now look further ahead.

## 2.4. HUMAN-LEVEL INTELLIGENCE?

The scenarios discussed in the last section are sufficiently near term that we need to plan for them and adjust to them. But what about the longer-term prospects? These are murkier, and there is no consensus among experts on the speed of advance in machine intelligence—and indeed on what the limits to AI might be. It seems plausible that an AI linked to the internet could 'clean up' on the stock market by analysing far more data far faster than any human. To some extent this is what quantitative hedge funds are already doing. But for interactions with humans, or even with the complex and fast-changing environment encountered by a driverless car on an ordinary road, processing power is not enough; computers would need sensors that enable them to see and hear as well as humans do, and the software to process and interpret what the sensors relay.

But even that would not be sufficient. Computers learn from a 'training set' of similar activities, where success is immediately 'rewarded' and reinforced. Game-playing computers play millions of games; photo-interpreting computers gain expertise by studying millions of images; for driverless cars to achieve this expertise, they would need to communicate with one another, to share and update their knowledge. But learning about human behaviour involves observing actual people in real homes or workplaces. The machine would feel sensorily deprived by the slowness of real life and would be bewildered. To quote Stuart Russell, a leading AI theorist, 'it could try all kinds of things: scrambling eggs, stacking wooden blocks, chewing wires, poking its finger into electric outlets. But nothing would produce a strong enough feedback loop to convince the computer it was on the right track and lead it to the next necessary action'.[12]

Only when this barrier can be surmounted will AIs truly be perceived as intelligent beings, to which (or to whom) we can relate, at least in some respects, as we do to other people. And their far faster 'thoughts' and reactions could then give them an advantage over us.

Some scientists fear that computers may develop 'minds of their own' and pursue goals hostile to humanity. Would a powerful futuristic AI remain docile, or 'go rogue'? Would it understand human goals and motives and align with them? Would it learn enough ethics and common sense so that it 'knew' when these should override its other motives? If it could infiltrate the internet of things, it could manipulate the rest of the world. Its goals may be contrary to human wishes, or it may even treat humans as encumbrances. AI must have a 'goal', but what really is difficult to instil is 'common sense'. AI should not pursue its goal obsessively and should be prepared to desist from its efforts rather than violating ethical norms.

Computers will vastly enhance mathematical skills, and perhaps even creativity. Already our smartphones substitute for routine memory storage and give near-instant access to the world's information. Soon translation between languages will be routine. The next step could be to 'plug in' extra memory or acquire language skills by direct input into the brain—though the feasibility of this isn't clear. If we can augment our brains with electronic implants, we might be able to download our thoughts and memories into a machine. If present technical trends proceed unimpeded, then some people now living could attain immortality—at least in the limited sense that their downloaded thoughts and memories could have a life span unconstrained by their present bodies. Those who seek this kind of eternal life will, in old-style spiritualist parlance, 'go over to the other side'.

We then confront the classic philosophical problem of personal identity. If your brain were downloaded into a machine, in what sense would it still be 'you'? Should you feel relaxed about your body then being destroyed? What would happen if several 'clones' were made of 'you'? And is the input into our sense organs, and physical interactions with the real external world, so essential to our being that this transition would be not only abhorrent but also impossible? These are ancient conundrums for philosophers, but practical ethicists may soon need to address them because they might be relevant to choices that real humans will make within this century.

In regard to all these post-2050 speculations, we don't know where the boundary lies between what may happen and what will remain science fiction—just as we don't know whether to take seriously Freeman Dyson's vision of biohacking by children. There are widely divergent views. Some experts, for instance Stuart Russell at Berkeley, and Demis Hassabis of DeepMind, think that the AI field, like synthetic biotech, already needs guidelines for 'responsible innovation'. Moreover, the fact that AlphaGo achieved a goal that its creators thought would have taken several more years to reach has rendered DeepMind's staff even more bullish about the speed of advancement. But others, like the roboticist Rodney Brooks (creator of the Baxter robot and the Roomba vacuum cleaner) think these concerns are too far from realisation to be worth worrying about—they remain less anxious about artificial intelligence than about real stupidity. Companies like Google, working closely with academia and government, lead the research into AI. These sectors now speak with one voice in highlighting the need to promote 'robust and beneficial' AI, but tensions may emerge when AI moves from the research and development phase to being a potentially massive money-spinner for global companies.

But does it matter if AI systems are having conscious thoughts in the sense that humans do? In the view of the computer science pioneer Edsger Dijkstra, it's a nonquestion: 'Whether machines can think is about as relevant as the question of whether submarines can swim'. Both a whale and a submarine make forward progress through the water, but they do it in fundamentally different ways. But to many it matters deeply whether intelligent machines are self-aware. In a scenario (see section 3.5) where future evolution is dominated by entities that are electronic, rather than having the 'wet' hardware we have in our skulls, it would seem depressing if we'd been surpassed in competence by 'zombies' who couldn't appreciate the wonders of the universe they were in and couldn't 'sense' the outside world as humans can. Be that as it may, society will be transformed by autonomous robots, even though the jury's out on whether they'll possess what we'd call real understanding or whether they'll be 'idiot savants'—with competence without comprehension.

A sufficiently versatile superintelligent robot could be the last invention that humans need to make. Once machines surpass human intelligence, they could design and assemble a new generation of even more intelligent machines. Some of the 'staples' of speculative science that flummox physicists today—time travel, space warps, and the ultracomplex—may be harnessed by the new machines, transforming the world physically. Ray Kurzweil (mentioned in section 2.1 in connection with cryonics) argues that this could lead to a runaway intelligence explosion: the 'singularity'.[13]

Few people doubt that machines will one day surpass most distinctively human capabilities; the disagreements are about the rate of travel, not the direction. If the AI enthusiasts are vindicated, it may take just decades before flesh-and-blood humans are transcended—or it may take centuries. But, compared to the aeons of evolutionary time that led to humanity's emergence, even that is a mere blink of the eye. This is not a fatalistic projection. It is cause for optimism. The civilisation that supplants us could accomplish unimaginable advances—feats, perhaps, that we cannot even understand. I'll scan horizons beyond the Earth in chapter 3.

## 2.5. TRULY EXISTENTIAL RISKS?

Our world increasingly depends on elaborate networks: electricity power grids, air traffic control, international finance, globally dispersed manu-facturing, and so forth. Unless these networks are highly resilient, their benefits could be outweighed by catastrophic (albeit rare) breakdowns—real-world analogues of what happened in the 2008 global financial crisis. Cities would be paralysed without electricity—the lights would go out, but that would be far from the most serious consequence. Within a few days our cities would be uninhabitable and an-archic. Air travel can spread a pandemic worldwide within days, wreaking havoc on the disorganised megacities of the developing world. And social media can spread panic and rumour, and economic contagion, literally at the speed of light.

When we realise the power of biotech, robot-ics, cybertechnology, and AI—and, still more, their potential in the coming decades—we can't avoid anxieties about how this empowerment could be misused. The historical record reveals episodes when 'civilisations' have crumbled and even been extinguished. Our world is so interconnected it's unlikely a catastrophe could hit any region without its consequences cascading globally. For the first time, we need to contemplate a collapse—societal or ecological—that would be a truly global setback to civilisation. The setback could be temporary. On the other hand, it could be so devastating (and could have entailed so much environmental or genetic degradation) that the survivors could never regenerate a civilisation at the present level.

But this prompts the question: could there be a separate class of extreme events that would be 'curtains' for us all—catastrophes that could snuff out all humanity or even all life? Physicists working on the Manhattan Project during World War II raised these kinds of Promethean concerns. Could we be absolutely sure that a nuclear explosion wouldn't ignite all the world's atmosphere or oceans? Before the 1945 Trinity Test of the first atomic bomb in New Mexico, Edward Teller and two colleagues addressed this issue in a calculation that was (much later) published by the Los Alamos Laboratory; they convinced themselves that there was a large safety factor. And luckily, they were right. We now know for certain that a single nuclear weapon, devastating though it is, cannot trigger a nuclear chain reaction that would utterly destroy the Earth or its atmosphere.

But what about even more extreme experiments? Physicists aim to understand the particles that make up the world and the forces that govern those particles. They are eager to probe the most extreme energies, pressures, and temperatures; for this purpose, they build huge, elaborate machines—particle accelerators. The optimum way to produce an intense concentration of energy is to accelerate atoms to enormous speeds, close to the speed of light, and crash them together. When two atoms crash together, their constituent protons and neutrons implode to a density and pressure far greater than when they were packed into a normal nucleus, releasing their constituent quarks. They may then break up into still smaller particles. The conditions replicate, in microcosm, those that prevailed in the first nanosecond after the big bang.

Some physicists raised the possibility that these experiments might do something far worse—destroy the Earth, or even the entire universe. Maybe a black hole could form, and then suck in everything around it. According to Einstein's theory of relativity, the energy needed to make even the smallest black hole would far exceed what these collisions could generate. Some new theories, however, invoke extra spatial dimensions beyond our usual three; a consequence would be to strengthen gravity's grip, rendering it less difficult for a small object to implode into a black hole.

The second scary possibility is that the quarks would reassemble themselves into compressed objects called strangelets. That in itself would be harmless. However, under some hypotheses, a strangelet could, by contagion, convert anything else it encountered into a new form of matter, transforming the entire Earth into a hyperdense sphere about a hundred metres across.

The third risk from these collision experiments is still more exotic, and potentially the most disastrous of all: a catastrophe that engulfs space itself. Empty space—what physicists call 'the vacuum'—is more than just nothingness. It is the arena for everything that happens; it has, latent in it, all the forces and particles that govern the physical world. It is the repository of the dark energy that controls the universe's fate. Space might exist in different 'phases', as water can exist in three forms: ice, liquid, or steam. Moreover, the present vacuum could be fragile and unstable. The analogy here is with water that is 'supercooled'. Water can cool below its normal freezing point if it is pure and still; however, it only takes a small localised disturbance—for instance, a speck of dust falling into it—to trigger supercooled water's conversion into ice. Likewise, some have speculated that the concentrated energy created when particles crash together could trigger a 'phase transition' that would rip the fabric of space. This would be a cosmic calamity—not just a terrestrial one.

The most favoured theories are reassuring; they imply that the risks from the kind of experiments within our current powers are zero. However, physicists can dream up alternative theories (and write down equations for them) that are consistent with everything we know, and therefore can't be absolutely ruled out, which would allow one or another of these catastrophes to happen. These alternative theories may not be frontrunners, but are they all so incredible that we needn't worry?

Physicists were (in my view quite rightly) pressured to address these speculative 'existential risks' when powerful new accelerators came on line at the Brookhaven National Laboratory and at CERN in Geneva, generating unprecedented concentrations of energy. Fortunately, reassurance could be offered; indeed, I was one of those who pointed out that 'cosmic rays'—particles of much higher energies than can be made in accelerators—collide frequently in the galaxy but haven't ripped space apart.[14] And they have penetrated very dense stars without triggering their conversion into strangelets.

So how risk averse should we be? Some would argue that odds of ten million to one against an existential disaster would be good enough, because that is below the chance that, within the next year, an asteroid large enough to cause global devastation will hit the Earth. (This is like arguing that the extra carcinogenic effect of artificial radiation is acceptable if it doesn't so much as double the risk from natural radiation—radon in the local rocks, for example.) But to some, this limit may not seem stringent enough. If there were a threat to the entire Earth, the public might properly demand assurance that the probability is below one in a billion—even one in a trillion—before sanctioning such an experiment if the purpose was simply to assuage the curiosity of theoretical physicists.

Can we credibly give such assurances? We may offer these odds against the Sun not rising tomorrow, or against a fair die giving one hundred sixes in a row, because we're confident that we understand these things. But if our understanding is shaky—as it plainly is at the frontiers of physics—we can't really assign a probability, or confidently assert that something is unlikely. It's presumptuous to place confidence in any theories about what happens when atoms are smashed together with unprecedented energy. If a congressional committee asked: 'Are you really claiming that there's less than a one in a billion chance that you're wrong?' I'd feel uncomfortable saying yes.

But on the other hand, if a congressman asked: 'Could such an experiment disclose a transformative discovery that—for instance—provided a new source of energy for the world?' I'd again offer odds against it. The issue is then the relative likelihood of these two unlikely events—one hugely beneficial; the other catastrophic. I would guess that the 'upside'—a benefit to humanity—though highly improbable, was much less unlikely than the 'universal doom' scenario. Such thoughts would remove any compunction about going ahead—but it is impossible to quantify the relative probabilities. So, it might be hard to make a convincingly reassuring case for such a Faustian bargain. Innovation is often hazardous, but if we don't take risks we may forgo benefits. Application of the 'precautionary principle' has an opportunity cost—'the hidden cost of saying no'.

Nonetheless, physicists should be circumspect about carrying out experiments that generate conditions with no precedent, even in the cosmos. In the same way, biologists should avoid creation of potentially devastating genetically modified pathogens, or large-scale modification of the human germ line. Cyberexperts are aware of the risk of a cascading breakdown in global infrastructure. Innovators who are furthering the beneficent uses of advanced AI should avoid scenarios where a machine 'takes over'. Many of us are inclined to dismiss these risks as science fiction—but given the stakes, they should not be ignored, even if deemed highly improbable.

These examples of near-existential risks also exemplify the need for interdisciplinary expertise, and for proper interaction between experts and the public. Moreover, ensuring that novel technologies are harnessed optimally will require communities to think globally and in a longer-term context. These ethical and political issues are discussed further in chapter 5.

And, by the way, the priority we should accord to avoiding truly existential disasters depends on an ethical question that has been discussed by the philosopher Derek Parfit: the rights of those who aren't yet born. Consider two scenarios: scenario *A* wipes out 90 percent of humanity; scenario *B* wipes out 100 percent. How much worse is *B* than *A*? Some would say 10 percent worse: the body count is 10 percent higher. But Parfit would argue that *B* might be *incomparably* worse, because human extinction forecloses the existence of billions, even trillions, of future people—and indeed an open-ended posthuman future spreading far beyond the Earth.[15] Some philosophers criticise Parfit's argument, denying that 'possible people' should be weighted as much as actual ones ('We want to make more people happy, not to make more happy people'). And even if one takes these naive utilitarian arguments seriously, one should note that if aliens already existed (see section 3.5), terrestrial expansion, by squeezing their habitats, might make a net negative contribution to overall 'cosmic contentment'!

However, aside from these intellectual games about 'possible people', the prospect of an end to the human story would sadden those of us now living. Most of us, aware of the heritage we've been left by past generations, would be depressed if we believed that there would not be many generations to come.

(This is a megaversion of the issues that arise in climate policy, discussed in section 1.5, where it is controversial how much weight we should give to those as yet unborn who will live a century from now. It also influences our attitude to global population growth.)

Even if we'd bet against an accelerator experiment or a genetic disaster destroying humanity, I think it is worth considering such scenarios as a 'thought experiment'. We have no grounds for assuming that human-induced threats far worse than those on our current risk register can be dismissed. Indeed, we have zero grounds for confidence that we can survive the worst that future technologies could bring. It's an important maxim that 'the unfamiliar is not the same as the improbable'.[16]

These ethical questions are far from the 'everyday', but it's not premature to address them—it's good that some philosophers are doing so. But they also challenge scientists. Indeed, they suggest an extra reason for addressing questions about the physical world that may seem arcane and remote: the stability of space itself, the emergence of life, and the extent and nature of what we might call 'physical reality'.

Such thoughts lead us from a terrestrial focus to a more cosmic perspective, which will be the theme of the next chapter. Despite the 'glamour' of human spaceflight, space is a hostile environment to which humans are ill-adapted. So, it's there that robots, enabled by human-level AI, will have the grandest scope, and where humans may use bio- and cyber-techniques to evolve further.

# 3

# HUMANITY IN A COSMIC PERSPECTIVE

## 3.1. THE EARTH IN A COSMIC CONTEXT

In 1968, the Apollo 8 astronaut Bill Anders photographed 'Earthrise', showing the distant Earth, shining above the lunar horizon. He didn't realise that it would become an iconic image for the global environmental movement. It revealed Earth's delicate biosphere, contrasted with the sterile moonscape where Neil Armstrong, one year later, took his 'one small step'. Another famous image was taken in 1990 by the probe *Voyager 1* from a distance of six billion kilometres. The Earth appeared as a 'pale blue dot', which inspired Carl Sagan's thoughts:[1]

> Look again at that dot. That's here. That's home. That's us. On it everyone you love, everyone you know, everyone you ever heard of, every human

being who ever was, lived out their lives. . . .
Every saint and sinner in the history of our species lived there—on a mote of dust suspended in
a sunbeam.

Our planet is a lonely speck in the great envelop-
ing cosmic dark. There is no hint that help will
come from elsewhere to save us from ourselves—
The Earth is the only world known so far to harbor
life. Like it or not, for the moment the Earth is
where we make our stand.

These sentiments resonate today; indeed, there
is serious discussion about how cosmic exploration
far beyond the solar system, by machines if not by
humans, could become reality—albeit in the remote
future. (*Voyager 1* is now, after more than forty years,
still in the outskirts of the solar system. It will take it
tens of thousands of years to reach the nearest star.)

We've been aware since Darwin of the Earth's long
history. He concludes *On the Origin of Species* with
these familiar words: 'Whilst this planet has gone
cycling on according to the fixed law of gravity, from
so simple a beginning endless forms most beauti-
ful and most wonderful have been, and are being,
evolved'. We now speculate about equally long time

spans stretching into the future, and these will be the themes of this chapter.

Darwin's 'simple beginning'—the young Earth—is complex in its chemistry and structure. Astronomers aim to probe still further back than Darwin and the geologists were able to—to the origin of planets, stars, and their constituent atoms.

Our entire solar system condensed from a swirling disc of dusty gas about 4.5 billion years ago. But where did the atoms come from—why are oxygen and iron atoms common, but not gold atoms? Darwin would not have fully understood this question; in his time, the very existence of atoms was controversial. But we now know that not only do we share a common origin, and many genes, with the entire web of life on Earth, but we also are linked to the cosmos. The Sun and stars are nuclear fusion reactors. They derive their power by fusing hydrogen into helium, and then helium into carbon, oxygen, phosphorus, and iron, and other elements in the periodic table. When stars end their lives, they expel 'processed' material back into interstellar space (via supernova explosions in the case of heavy stars). Some material is then recycled into new stars. The Sun was one such star.

A typical carbon atom, in one of the trillions of $CO_2$ molecules that we inhale with each breath, has an eventful history stretching back more than five billion years. The atom was perhaps released into the atmosphere when a lump of coal was burned—a lump that was itself the remnant of a tree in a primeval forest two hundred million years ago—and before that had been cycled between the Earth's crust, biosphere, and oceans ever since our planet's formation. Tracing back further we would find that the atom was forged in an ancient star that exploded, ejecting carbon atoms that wandered in interstellar space, condensing into a proto–solar system and thence into the young Earth. We are literally the ashes of long-dead stars—or (less romantically) the nuclear waste from the fuel that made stars shine.

Astronomy is an ancient science—perhaps the oldest apart from medicine (and I'd argue the first to do more good than harm—by improving the calendar, timekeeping, and navigation). And for the last few decades cosmic exploration has been on a roll. There are human footprints on the Moon. Robotic probes to other planets have beamed back pictures of fascinating and varied worlds—and landed on some of them. Modern telescopes have enlarged our cosmic horizons. And these telescopes have revealed a 'zoo' of extraordinary objects—black holes, neutron stars, and colossal explosions. Our Sun is embedded within our galaxy, the Milky Way, which contains more than a hundred billion stars, all orbiting around a central hub where lurks a massive black hole. And this is just one of one hundred billion galaxies visible through the telescopes. We've even detected 'echoes' of the 'big bang' that triggered our entire expanding universe 13.8 billion years ago. This is how the universe was born—and with it, all the basic particles of nature.

Armchair theorists like myself can claim little credit for this progress; it is owed mainly to improvements in telescopes, spacecraft, and computers. Thanks to these advances we're starting to understand the chain of events whereby, from a mysterious beginning when everything was squeezed to immense temperatures and densities, atoms, stars, galaxies, and planets emerged—and how on one planet, Earth, atoms assembled into the first living things, initiating the Darwinian evolution that's led to creatures like us, able to ponder the mystery of it all.

Science is a truly global culture—spanning all boundaries of nationality and faith. That's especially true of astronomy. The night sky is the most universal feature of our environment. Throughout human history, people all over the world have gazed at the stars—interpreting them in different ways. Just within the last decade the night sky has become vastly more interesting than it was to our ancestors. We've learned that most stars aren't just twinkling points of light but are orbited by planets, just as the Sun is. Amazingly, our galaxy harbours many millions of planets like the Earth—planets that seem habitable. But are they actually inhabited—is there life, even intelligent life, out there? It's hard to imagine a question of greater importance for understanding our place in the cosmic scheme of things.

It is clear from the extensive media coverage that these issues fascinate millions. It's gratifying to astronomers (and to those in fields like ecology) that their fields attract such broad popular interest. I'd derive far less satisfaction from my research if I could only discuss it with a few fellow specialists. Moreover, the subject has a positive and nonthreatening image—in contrast to the public ambivalence about, for instance, nuclear science, robotics, or genetics.

If I'm on a plane and don't want to chat with the person in the next seat, a sure conversation stopper is to say 'I'm a mathematician'. In contrast, saying 'I'm an astronomer' often stimulates interest. And the number one inquiry is then usually 'do you believe in aliens, or are we alone?' The topic fascinates me too, so I'm always glad to discuss it. And it has another virtue as a conversation starter. Nobody yet knows the answer, so there is less of a barrier between the 'expert' and the general inquirer. There's nothing new about this fascination; but now, for the first time, we have hope of an answer.

Speculations on 'the plurality of inhabited worlds' date back to antiquity. From the seventeenth to the nineteenth century, it was widely suspected that the other planets of the solar system were inhabited. The reasoning was more often theological than scientific. Eminent nineteenth-century thinkers argued that life must pervade the cosmos, because otherwise such vast domains of space would seem such a waste of the Creator's efforts. An amusing critique of such ideas is given in the impressive book *Man's Place in the Universe* by Alfred Russel Wallace, the codeveloper of the theory of natural selection.[2] Wallace is especially scathing about the physicist David Brewster (remembered by physicists for the 'Brewster angle' in optics), who conjectured on such grounds that even the Moon must be inhabited. Brewster argued in his book *More Worlds Than One* that had the Moon 'been destined to be merely a lamp to our Earth, there was no occasion to variegate its surface with lofty mountains and extinct volcanoes and cover it with large patches of matter that reflect different quantities of light and give its surface the appearance of continents and seas. It would have been a better lamp had it been a smooth piece of lime or of chalk'.

By the end of the nineteenth century, many astronomers were so convinced that life existed on other planets in the solar system that a prize of one hundred thousand francs was offered to the first person to make contact. And the prize specifically excluded contact with Martians—that was considered far too easy! The erroneous claim that Mars was crisscrossed by canals had been taken as proof positive of intelligent life on the red planet.

The space age brought sobering news. Venus, a cloudy planet that promised a lush tropical swamp-world, turned out to be a crushing, caustic hellhole. Mercury was a pockmarked blistering rock. Even Mars, the most Earthlike planet, is now revealed as a frigid desert with a very thin atmosphere. NASA's *Curiosity* probe may, however, have found water.

And it detected methane gas burping from below the surface—perhaps from rotting organisms that lived long ago—though there seems no interesting life there now.

In the still-colder objects farther from the Sun, the smart money would be on Europa, one of Jupiter's moons, and Enceladus, a moon of Saturn. These are covered in ice, and there could be creatures swimming in the oceans beneath; space probes are being planned that will search for them. And there could be exotic life in the methane lakes of Titan, another of Saturn's moons. But nobody can be optimistic.

Within the solar system, Earth is the Goldilocks planet—not too hot and not too cold. Were it too hot, even the most tenacious life would fry. But if it were too cold, the processes that created and nourished life would have happened far too slowly. The discovery of even vestigial life-forms elsewhere in the solar system would be of epochal importance. That's because it would tell us that life wasn't a rare fluke but was widespread in the cosmos. At the moment we know of only one place—Earth—where life began. It is logically possible (indeed, some argue that it's plausible) that life's origin requires such special contingencies that it only happened once in our entire galaxy. But if it arose twice within a single planetary system, then it must be common.

(There is one important proviso: before drawing this inference about life's ubiquity we must be sure that two life-forms emerged independently rather than being transported from one location to another. For that reason, life under Europa's ice would clinch the case more than life on Mars, because it's conceivable that we all have Martian ancestry—having evolved from primitive life carried on a rock shot off Mars by an asteroid impact and propelled towards Earth.)

## 3.2. BEYOND OUR SOLAR SYSTEM

To find promising 'real estate' on which life can exist, we must extend our gaze beyond our solar system— beyond the reach of any probe we can devise today. What has transformed and energised the whole field of exobiology is the realisation that most stars are orbited by planets. The Italian monk Giordano Bruno speculated about this in the sixteenth century. From the 1940s onward, astronomers suspected he was correct. An earlier theory that the solar system formed from a filament torn out of the Sun by the gravitational pull of a close-passing star (which would have implied that planetary systems were rare) had by then been discredited. This theory was superseded by the idea that when an interstellar cloud contracted under gravity to form a star, it would, if it were rotating, 'spin off' a disc whose constituent gas and dust would agglomerate into planets. But it wasn't until the 1990s that evidence for exoplanets started to emerge. Most exoplanets are not detected directly; they are inferred through careful observation of the star they're orbiting. There are two main techniques.

The first is this. If a star is orbited by a planet, then both planet and star move around their centre of mass—what's called the barycentre. The star, being more massive, moves slower. But the cyclic motion induced by an orbiting planet can be detected by precise study of the starlight, which reveals a changing Doppler effect. The first success came in 1995 when Michel Mayor and Didier Queloz, based at the Observatory of Geneva, found a 'Jupiter-mass' planet around the nearby star 51 Pegasi.[3] In the subsequent years, more than four hundred exoplanets have been found in this way. This 'stellar wobble' technique pertains mainly to 'giant' planets—objects the size of Saturn or Jupiter.

Possible 'twins' of Earth are specially interesting: planets the same size as ours, orbiting other Sun-like stars, on orbits with temperatures such that water neither boils nor stays frozen. But detecting these— hundreds of times less massive than Jupiter—is a real challenge. They induce wobbles of merely centimetres per second in their parent star—this motion has hitherto been too small for the Doppler method to detect (though the instrumentation advances apace).

But there's a second technique: we can look for the planets' shadows. A star would appear to dim slightly when a planet was 'in transit' in front of it; these dimmings would repeat at regular intervals. Such data reveal two things: the interval between successive dimmings tells us the length of the planet's year, and the amplitude of the dimming tells us what fraction of the star's light a planet blocks out during the transit, and therefore how big it is.

The most important search (so far) for transiting planets was carried out by a NASA spacecraft named after astronomer Johannes Kepler,[4] which spent more than three years measuring the brightness of 150,000 stars, to a precision of one part in 100,000— it did this once or more times an hour for each star. *Kepler* found thousands of transiting planets, some no bigger than Earth. The prime mover behind the *Kepler* project was Bill Borucki, an American engineer who had worked for NASA since 1964. He conceived the concept in the 1980s and doggedly pursued it despite funding setbacks and initial scepticism from many in the community of 'established' astronomers. His triumphant success—achieved when he was already in his seventies—deserves special acclaim. It reminds us of how much even the 'purest' science owes to the instrument builders.

There is variety among the already discovered exoplanets. Some are on eccentric orbits. And one planet has four suns in its sky; it is orbiting a binary star, which is orbited by another binary star. This discovery involved amateur 'planet hunters'; any enthusiast had the chance to access *Kepler* data from some stars, and the human eye was able to pick out 'dips' in the stars' brightness (which occurred less regularly than in the case when a planet orbits a single star).

There's a planet orbiting the nearest star, Proxima Centauri, which is only four light years from Earth. Proxima Centauri is a so-called M dwarf star, about a hundred times fainter than our Sun. In 2017 a team led by the Belgian astronomer Michaël Gillon discovered a miniature solar system around another

M dwarf;[5] seven planets, with 'years' lasting from 1.5 to 18.8 Earth days, are orbiting around it. The outer three are in the habitable zone. They'd be spectacular places to live. Viewed from the surface of one of the planets, the others would swing fast across the sky, looming as large as our Moon does to us. But they're very un-Earthly. They're probably tidally locked so that they present the same face to their star—one hemisphere in perpetual light; the other always dark. (In the unlikely event that it harboured intelligent life, a kind of 'segregation' might prevail—the astronomers quarantined in one hemisphere, everyone else in the other!) But it's likely that their atmospheres have been stripped away by the intense magnetic flaring that is common on M dwarf stars, rendering them less propitious for life.

The known exoplanets are nearly all inferred indirectly, by detecting their effect on the motions or brightness of the stars they're orbiting. We'd really like to see them directly but that's hard. To realise just how hard, suppose that aliens existed, and that an alien astronomer with a powerful telescope was viewing the Earth from (say) thirty light years away—the distance of a nearby star. Our planet would seem, in Carl Sagan's phrase, a 'pale blue dot', very close to a star (our Sun) that outshines it by many billions: a firefly next to a searchlight. The shade of blue would be slightly different, depending on whether the Pacific Ocean or the Eurasian land mass was facing them. The alien astronomers could infer the length of our day, the seasons, the existence of continents and oceans, and the climate. By analysing the faint light, the astronomers could infer that the Earth had a green surface and an oxygenated atmosphere.

Today, the largest terrestrial telescopes are built by international consortia. They are mushrooming on Mauna Kea (Hawai'i) and under the clear dry skies of the high Andes in Chile. And South Africa not only has one of the world's largest optical telescopes but will also have a leadership role, along with Australia, in constructing the world's largest radio telescope, the Square Kilometre Array. A telescope now being built on a Chilean mountaintop by European astronomers will have the required sensitivity to pick up light from planets the same size as Earth orbiting other sun-like stars. It's called the European Extremely Large Telescope (E-ELT)—literal rather than imaginative nomenclature! Newton's first reflecting telescope had a 10-centimeter-diameter mirror; the E-ELT will be 39 meters—a mosaic of small mirrors with a total collecting area more than a hundred thousand times larger.

From the statistics of planets around the nearby stars studied so far, we can infer that the entire Milky Way harbours around a billion planets that are 'Earthlike' in the sense that they are about the size of Earth and at a distance from their parent star such that water can exist, neither boiling away nor staying permanently frozen. We'd expect a variety: some might be 'waterworlds', completely covered with oceans; others might (like Venus) have been heated and sterilised by an extreme 'greenhouse effect'.

How many of these planets might harbour lifeforms far more interesting and exotic than anything we might find on Mars—even something that could be called intelligent? We don't know what the odds are. Indeed, we can't yet exclude the possibility that life's origin—the emergence, from a chemical 'mix', of a metabolising and reproducing entity—involved a fluke so rare that it happened only once in our entire galaxy. On the other hand, this crucial transition might have been almost inevitable given the 'right' environment. We just don't know—nor do we know if the DNA/RNA chemistry of terrestrial life is the only possibility, or just one chemical basis among many options that could be realised elsewhere. Nor, even more fundamentally, do we know whether liquid water really is crucial. If there were a chemical path whereby life could emerge in the cold methane lakes of Titan, our definition of 'habitable planets' would be very much broader.

These key issues may soon be clarified. The origin of life is now attracting stronger interest; it's no longer deemed to be one of those ultrachallenging problems (consciousness, for instance, is still in this category) which, though manifestly important, don't seem timely or tractable—and are relegated to the 'too difficult' box. Understanding life's beginnings is important not only for assessing the likelihood of alien life but also because life's emergence on Earth is still a mystery.

We should be open-minded about where in the cosmos life might emerge and what forms it could take—and devote some thought to non-Earthlike life in non-Earthlike locations. Even here on Earth, life survives in the most inhospitable places—in black caves where sunlight has been blocked for thousands of years, inside arid desert rocks, deep underground, and around hot vents in the deepest ocean bed. But it makes sense to start with what we know (the 'searching under the streetlight' strategy) and to deploy all available techniques to discover whether any Earthlike exoplanet atmospheres display evidence for a biosphere. Clues should come, in the next decade or two, from the deep space James Webb Space Telescope and from the E-ELT and similar giant telescopes on the ground that will come on line in the 2020s.

Even these next-generation telescopes will have a hard job separating out the spectrum of the planet's atmosphere from the spectrum of the brighter central star. But, looking beyond midcentury, one can imagine an array of vast space telescopes, each with gossamer-thin kilometre-scale mirrors, being assembled in deep space by robotic fabricators. By 2068, the centenary of the Apollo 8 'Earthrise' photo, such an instrument could give us an even more inspirational image: another Earth orbiting a distant star.

## 3.3. SPACEFLIGHT—MANNED AND UNMANNED

Among my favourite things to read during my childhood (in England, way back in the 1950s), was a comic called the Eagle, especially the adventures of 'Dan Dare—Pilot of the Future'—where the brilliant artwork depicted orbiting cities, jet packs, and alien invaders. When spaceflight became real, the suits worn by NASA astronauts (and their Soviet 'cosmonaut' counterparts) were therefore familiar, as were the routines of launching, docking, and so forth. My generation avidly followed the succession of heroic pioneering exploits: Yuri Gagarin's first orbital flight, Alexey Leonov's first space walk, and then, of course, the lunar landings. I recall a visit to my home town by John Glenn, the first American to go into orbit. He was asked what he was thinking while in the rocket's nose cone, awaiting launch. He responded, 'I was thinking that there were twenty thousand parts in this rocket, and each was made by the lowest bidder'. (Glenn later became a US senator, and, later still, the oldest astronaut when, at age seventy-seven, he became part of the STS-95 Space Shuttle crew.)

Only twelve years elapsed between the flight of the Soviet *Sputnik 1*—the first artificial object to go into orbit—and the historic 'one small step' on the lunar surface in 1969. I never look at the Moon without being reminded of Neil Armstrong and Buzz Aldrin. Their exploits seem even more heroic in retrospect, when we realise how they depended on primitive computing and untested equipment. Indeed,

President Nixon's speechwriter William Safire had drafted a eulogy to be given if the astronauts had crash-landed on the Moon or were stranded there:

> Fate has ordained that the men who went to the moon to explore in peace will stay on the moon to rest in peace. [They] know that there is no hope for their recovery. But they also know that there is hope for mankind in their sacrifice.

The Apollo programme remains, a half century later, the high point of human ventures into space. It was a 'space race' against the Russians—a contest in superpower rivalry. Had that momentum been maintained, there would surely be footprints on Mars by now; that's what our generation expected. However, once that race was won, there was no motivation for continuing the requisite expenditure. In the 1960s, NASA absorbed more than 4 percent of the US federal budget. The current figure is 0.6 percent. Today's young people know Americans landed men on the Moon. They know the Egyptians built pyramids. But these enterprises seem like ancient history, motivated by almost equally bizarre national goals.

Hundreds more have ventured into space in the ensuing decades—but, anticlimactically, they have done no more than circle the Earth in low orbit. The International Space Station (ISS) was probably the most expensive artefact ever constructed. Its cost, plus that of the shuttles whose main purpose was to service it (though they have now been decommissioned) ran well into twelve figures. The scientific and technical payoff from the ISS hasn't been negligible, but it has been less cost effective than unmanned missions. Nor are these voyages inspiring in the way that the pioneering Russian and US space exploits were. The ISS only makes news when something goes wrong: when the loo fails, for instance; or when astronauts perform 'stunts', such as the Canadian Chris Hadfield's guitar playing and singing.

The hiatus in manned space exploration exemplifies that when there's no economic or political demand, what is actually done is far less than what could be achieved. (Supersonic flight is another example—the Concorde airliner went the way of the dinosaurs. In contrast, the spin-offs from IT have advanced, and spread globally, far faster than forecasters and management gurus predicted.)

Space technology has nonetheless burgeoned in the last four decades. We depend routinely on orbiting satellites for communication, satnav, environmental monitoring, surveillance, and weather forecasting. These services mainly use spacecraft that, though unmanned, are expensive and elaborate. But there is a growing market for relatively inexpensive miniaturised satellites, the demand for which several private companies are aiming to meet.

The San Francisco–based company PlanetLab has developed and launched swarms of shoebox-sized spacecraft with the collective mission of giving repeated imaging and global coverage, albeit at not-specially-sharp resolution (3–5 metres): the mantra (with only slight exaggeration) is to observe every tree in the world every day. Eighty-eight of the craft were launched in 2017 as payload on a single Indian rocket; Russian and US rockets have been used to launch more, as well as a fleet of somewhat larger and more elaborately equipped SkySats (each weighing 100 kilograms). For much sharper resolution, a larger satellite with more elaborate optics is needed, but there is nonetheless a commercial market for the data from these tiny 'cubesats' to monitor crops, construction sites, fishing boats, and suchlike; they are also useful for planning a response to disasters. Even smaller wafer-thin satellites can now be deployed—exploiting the technology that has emerged from the colossal investment in consumer microelectronics.

Telescopes in space offer astronomy a huge boost. Orbiting far above the blurring and absorptive effects of Earth's atmosphere, they have beamed back sharp images from the remotest parts of the cosmos. They have surveyed the sky in infrared, UV, X-ray, and gamma ray bands that don't penetrate the atmosphere and therefore can't be observed from the ground. They have revealed evidence for black holes and other exotica and have probed with high precision the 'afterglow of creation'—the microwaves pervading all space whose properties hold clues to the very beginning, when the entire observable cosmos was squeezed to microscopic size.

Of more immediate public appeal are the findings from spacecraft that have journeyed to all the planets of the solar system. NASA's *New Horizons* beamed back amazing pictures from Pluto, ten thousand times farther away than the Moon. And the European Space Agency's *Rosetta* landed a robot on a comet. These spacecraft took five years to design and build and then nearly ten years journeying to their remote targets. The *Cassini* probe spent thirteen years studying Saturn and its moons and was even more venerable; more than twenty years elapsed between its launch and its final plunge into Saturn in late 2017. It is not hard to envisage how much more sophisticated today's follow-ups to these missions could be.

During this century, the entire solar system—planets, moons, and asteroids—will be explored and mapped by fleets of tiny robotic space probes, interacting with each other like a flock of birds. Giant robotic fabricators will be able to construct, in space, solar energy collectors and other objects. The Hubble telescope's successors, with oversize mirrors assembled in zero gravity, will further expand our vision of exoplanets, stars, galaxies, and the wider cosmos. The next step would be space mining and fabrication.

But will there be a role for humans? There's no denying that NASA's *Curiosity*, a vehicle the size of a small car that has since 2011 been trundling across a giant Martian crater, may miss startling discoveries that no human geologist could overlook. But machine learning is advancing fast, as is sensor technology. In contrast, the cost gap between manned and unmanned missions remains outsized. The practical case for manned spaceflight gets ever weaker with each advance in robots and miniaturisation.

If there were a revival of the 'Apollo spirit' and a renewed urge to build on its legacy, a permanently manned lunar base would be a credible next step. Its construction could be accomplished by robots—bringing supplies from Earth and mining some from the Moon. An especially propitious site is the Shackleton crater, at the lunar south pole, 21 kilometres across and with a rim 4 kilometres high. Because of the crater's location, its rim is always in sunlight and so escapes the extreme monthly temperature contrasts experienced on almost all the Moon's surface. Moreover, there may be a lot of ice in the crater's perpetually dark interior—crucial, of course, for sustaining a 'colony'.

It would make sense to build mainly on the half of the Moon that faces the Earth. But there is one exception: astronomers would like a giant telescope on the far side because it would then be shielded from the artificial emission from the Earth—offering a great advantage to radio astronomers seeking to detect very faint cosmic emissions.

NASA's manned space programme, ever since Apollo, has been constrained by public and political pressure to be risk-averse. The space shuttle failed twice in 135 launches. Astronauts or test pilots would willingly accept this level of risk—less than 2 percent. But the shuttle had, unwisely, been promoted as a safe vehicle for civilians (and a female schoolteacher, Christa McAuliffe, in the NASA Teacher in Space Project, was one of the casualties of the *Challenger* disaster). Each failure caused a national trauma in the United States and was followed by a hiatus while costly efforts were made (with very limited effect) to reduce risks still further.

I hope some people now living will walk on Mars—as an adventure, and as a step towards the stars. But NASA will confront political obstacles in achieving this goal within a feasible budget. China has the resources, the dirigiste government, and maybe the willingness to undertake an Apollo-style programme. If it wanted to assert its superpower status by a 'space spectacular' and to proclaim parity, China would need to leapfrog, rather than just rerun, what the United States had achieved fifty years earlier. It already plans a 'first' by landing on the far side of the Moon. A clearer-cut 'great leap forward' would involve footprints on Mars, not just on the Moon.

Leaving aside the Chinese, I think the future of manned spaceflight lies with privately funded adventurers, prepared to participate in a cut-price programme far riskier than western nations could impose on publicly supported civilians. SpaceX, led by Elon Musk (who also builds Tesla electric cars), or the rival effort, Blue Origin, bankrolled by Jeff Bezos, founder of Amazon, have berthed craft at the space station and will soon offer orbital flights to paying customers. These ventures—bringing a Silicon Valley culture into a domain long dominated by NASA and a few aerospace conglomerates— have shown it's possible to recover and reuse the launch rocket's first stage—presaging real cost savings. They have innovated and improved rocketry far faster than NASA or ESA has done—a SpaceX Falcon rocket is able to put a fifty-ton payload into orbit. The future role of the national agencies will be attenuated—becoming more akin to an airport than to an airline.

If I were an American, I would not support NASA's manned programme—I would argue that inspirationally led private companies should 'front' all manned missions as cut-price high-risk ventures. There would still be many volunteers—some perhaps even accepting 'one-way tickets'—driven by the same motives as early explorers, mountaineers, and the like. Indeed, it is time to eschew the mind-set that space ventures should be national (even international) projects—along with pretentious rhetoric where the word 'we' is used to denote the whole of humanity. There are some endeavours—tackling climate change, for instance—that can't be done without concerted international action. The exploitation of space need not be of this nature; it may need some public regulation, but the impetus can be private or corporate.

There are plans for week-long trips round the far side of the Moon—voyaging farther from Earth than anyone has gone before (but avoiding the greater challenge of a Moon landing and blast-off). A ticket has been sold (I'm told) for the second such flight but not the first. And Dennis Tito, an entrepreneur and former astronaut, has proposed, when a new heavy-lift launcher is available, to send people to Mars and back—without landing. This would require five hundred days in isolated confinement. The ideal crew would be a stable middle-aged couple—old enough to not be bothered about the high dose of radiation accumulated on the trip.

The phrase *space tourism* should be avoided. It lulls people into believing that such ventures are routine and low risk. And if that's the perception, the inevitable accidents will be as traumatic as those of the space shuttle. These exploits must be 'sold' as dangerous sports, or intrepid exploration.

The most crucial impediment to space flight, in Earth's orbit and for those venturing farther, stems from the intrinsic inefficiency of chemical fuel and the consequent requirement for launchers to carry a weight of fuel far exceeding that of the payload. So long as we are dependent on chemical fuels, interplanetary travel will remain a challenge. Nuclear power could be transformative. By allowing much higher in-course speeds, it would drastically cut the transit times to Mars or the asteroids (reducing not only astronauts' boredom but also their exposure to damaging radiation).

Greater efficiency would be achieved if the fuel supply could be on the ground and not carried into space. For instance, it might be technically possible to propel spacecraft into orbit via a 'space elevator'— a carbon-fibre rope 30,000 kilometres long anchored to the Earth (and powered from the ground), extending vertically up beyond the distance of a geostationary orbit so that it is held taut by centrifugal forces. An alternative scheme envisages a powerful laser beam generated on Earth that pushes on a 'sail' attached to the spacecraft; this might be feasible for lightweight space probes and could in principle accelerate them to 20 percent of the speed of light.[6]

Incidentally, more efficient on-board fuel could transform manned spaceflight from a high-precision to an almost unskilled operation. Driving a car would be a difficult enterprise if, as at present for space voyages, one had to programme the entire journey in detail beforehand, with minimal opportunities for steering along the way. If there were an abundance of fuel for midcourse corrections (and to brake and accelerate at will), then interplanetary navigation would be a low-skill task—simpler, even, than steering a car or ship, in that the destination is always in clear view.

By 2100 thrill seekers in the mould of (say) Felix Baumgartner (the Austrian skydiver who in 2012 broke the sound barrier in free fall from a high-altitude balloon) may have established 'bases' independent from the Earth—on Mars, or maybe on asteroids. Elon Musk (born in 1971) of SpaceX says he wants to die on Mars—but not on impact. But don't ever expect mass emigration from Earth. And here I disagree strongly with Musk and with my late Cambridge colleague Stephen Hawking, who enthuse about rapid build-up of large-scale Martian communities. It's a dangerous delusion to think that space offers an escape from Earth's problems. We've got to solve these problems here. Coping with climate change may seem daunting, but it's a doddle compared to terraforming Mars. No place in our solar system offers an environment even as clement as the Antarctic or the top of Everest. There's no 'Planet B' for ordinary risk-averse people.

But we (and our progeny here on Earth) should cheer on the brave space adventurers, because they will have a pivotal role in spearheading the post-human future and determining what happens in the twenty-second century and beyond.

## 3.4. TOWARDS A POST-HUMAN ERA?

Why will these space adventurers be so important? The space environment is inherently hostile for humans. So, because they will be ill-adapted to their new habitat, the pioneer explorers will have a more compelling incentive than those of us on Earth to redesign themselves. They'll harness the superpowerful genetic and cyborg technologies that will be developed in coming decades. These techniques will be, one hopes, heavily regulated on Earth, on prudential and ethical grounds, but 'settlers' on Mars will be far beyond the clutches of the regulators. We should wish them good luck in modifying their progeny to adapt to alien environments. This might be the first step towards divergence into a new species. Genetic modification would be supplemented by cyborg technology—indeed there may be a transition to fully inorganic intelligences. So, it's these space-faring adventurers, not those of us comfortably adapted to life on Earth, who will spearhead the posthuman era.

Before setting out from Earth, space voyagers, whatever their destination, would know what to expect at journey's end; robotic probes would have preceded them. The European explorers in earlier centuries who ventured across the Pacific went into the unknown to a far greater extent than any future explorers would (and faced more terrifying dangers)—there were no precursor expeditions to make maps, as there would be for space ventures. Future space-farers will always be able to communicate with Earth (albeit with a time lag). If precursor probes have revealed that there are wonders to explore, there will be a compelling motive—just as Captain Cook was incentivised by the biodiversity and beauties of the Pacific islands. But if there is nothing but sterility out there, the voyages might be better left to robotic fabricators.

Organic creatures need a planetary surface environment, but if posthumans make the transition to fully inorganic intelligences, they won't need an atmosphere. And they may prefer zero-g, especially for constructing extensive but lightweight habitats. So it's in deep space—not on Earth, or even on Mars— that nonbiological 'brains' may develop powers that humans can't even imagine. The timescales for technological advance are but an instant compared to the timescales of the Darwinian natural selection that led to humanity's emergence—and (more relevantly) they are less than a millionth of the vast expanses of cosmic time lying ahead. The outcomes of future technological evolution could surpass humans by as much as we (intellectually) surpass slime mould.

It's likely that 'inorganics'—intelligent electronic robots—will eventually gain dominance. This is because there are chemical and metabolic limits to the size and processing power of 'wet' organic brains. Maybe we're close to these already. But no such limits constrain electronic computers (still less, perhaps, quantum computers). So, by any definition of 'thinking', the amount and intensity that's done by organic human-type brains will be utterly swamped by the cerebrations of AI. We are perhaps near the end of Darwinian evolution, but a faster process, artificially directed enhancement of intelligence, is only just beginning. It will happen fastest away from the Earth—I wouldn't expect (and certainly wouldn't wish for) such rapid changes in humanity here on Earth though our survival will depend on ensuring that the AI on Earth remains 'benevolent'.

Philosophers debate whether 'consciousness' is special to the organic brains of humans, apes, and dogs. Might it be that robots, even if their intellects seem superhuman, will still lack self-awareness or inner life? The answer to this question crucially affects how we react to their 'takeover'. If the machines are zombies, we would not accord their experiences the same value as ours, and the posthuman future would seem bleak. But if they are conscious, why should we not welcome the prospect of their future hegemony?

The scenarios I've just described would have the consequence—a boost to human self-esteem—that even if life had originated only on the Earth, it need not remain a trivial feature of the cosmos; humans may be closer to the beginning than to the end of a process whereby ever more complex intelligence spreads through the galaxy. The leap to neighbouring stars is just an early step in this process. Interstellar voyages—or even intergalactic voyages—would hold no terrors for near-immortals.

Even though we are not the terminal branch of an evolutionary tree, we humans could claim truly cosmic significance for jump-starting the transition to electronic (and potentially immortal) entities, spreading their influence far beyond the Earth, and far transcending our limitations.

But the motives and the ethical constraints will then depend on the answer to one great astronomical question: Is there life—intelligent life—out there already?

## 3.5. ALIEN INTELLIGENCE?

Firm evidence for vegetation, primitive bugs, or bacteria on an exoplanet would be significant. But the thing that really fuels popular imagination is the prospect of advanced life—the 'aliens' familiar from science fiction.[7]

Even if primitive life were common, 'advanced' life may not be—its emergence may depend on many contingencies. The course of evolution on Earth was influenced by phases of glaciation, our planet's tectonic history, asteroid impacts, and so forth. Several authors have speculated about evolutionary 'bottlenecks'—key stages that are hard to transit. Perhaps the transition to multicellular life (which took two billion years on Earth) is one of these. Or the 'bottleneck' could come later. If, for instance, the dinosaurs hadn't been wiped out, the chain of mammalian evolution that led to humans may have been foreclosed; we can't predict whether another species would have taken our role. Some evolutionists regard the emergence of intelligence as an unlikely contingency, even in a complex biosphere.

Perhaps, more ominously, there could be a 'bottleneck' at our own evolutionary stage—the stage we're at during this century, when intelligent life develops powerful technology. The long-term prognosis for 'Earth-sourced' life depends on whether humans survive this phase—despite vulnerability to the kinds of hazards I've addressed in earlier chapters. This does not require that no terminal catastrophe ever befalls the Earth—only that, before that happens, some humans or artefacts have spread beyond their home planet.

As I've emphasised, we know too little about how life emerged to be able to say whether alien intelligence is likely or not. The cosmos could be teeming with varieties of complex life; if so, we could aspire to be minor members of a 'galactic club'. On the other hand, the emergence of intelligence may require such a rare chain of events—like winning a lottery—that it has not occurred anywhere else. That will disappoint those searching for aliens but would imply that our Earth could be the most important place in the galaxy, and that its future is of cosmic consequence.

It would plainly be a momentous discovery to detect any cosmic 'signal' that was manifestly artificial—radio 'beeps', or flashes of light from some celestial laser scanning the Earth. Searches for extraterrestrial intelligence (SETI) are worthwhile, even if the odds seem stacked against success, because the stakes are so high. Earlier searches led by Frank Drake, Carl Sagan, Nikolai Kardashev, and others didn't find anything artificial. But they were very limited—it's like claiming that there's no life in the oceans after analysing one glassful of seawater. That's why we should welcome the launch of Breakthrough Listen, a ten-year commitment by Yuri Milner, a Russian investor, to buy time on the world's best radio telescopes and develop instruments to scan the sky in a more comprehensive and sustained fashion than before. The searches will cover a wide range of radio and microwave frequencies, using specially developed signal processing equipment. And they will be supplemented by searches for 'flashes' of visible light or X-rays that don't seem to have a natural origin. Moreover, the advent of social media and citizen science will enable a global community of enthusiasts to download data and participate in this cosmic quest.

In popular culture, aliens are depicted as vaguely humanoid—generally two-legged, though maybe with tentacles, or eyes on stalks. Perhaps such creatures exist. But they aren't the kind of alien that we'd be most likely to detect. I would argue strongly that an ET transmission, if we were to find it, would more likely come from immensely intricate and powerful electronic brains. I infer this from what has happened on Earth, and—more important—how we expect life and intelligence to evolve in the far future. The first tiny organisms emerged when the Earth was young, nearly four billion years ago; this primordial biosphere has evolved into today's marvellously complex web of life—of which we humans are a part. But humans aren't the end of this process—indeed, they may not be even the halfway stage. So future evolution—the posthuman era, where the dominant creatures aren't flesh and blood—could extend billions of years into the future.

Suppose that there are many other planets where life began, and that on some of them Darwinian evolution followed a similar track to what has happened here. Even then, it's highly unlikely that the key stages would be synchronised. If the emergence of intelligence and technology on a planet lagged significantly behind what has happened on Earth (because the planet is younger, or because the 'bottlenecks' have taken longer to negotiate), then that planet would reveal no evidence of ET. But around a star older than the Sun, life could have had a head start of a billion years or more.

The history of human technological civilisation is measured in millennia (at most)—and it may be only one or two more centuries before humans are overtaken or transcended by inorganic intelligence, which will then persist, continuing to evolve, for billions of years. If 'organic' human-level intelligence is, generically, just a brief interlude before the machines take over, we would be most unlikely to 'catch' alien intelligence in the brief sliver of time when it was still in organic form. Were we to detect ET, it would be far more likely to be electronic.

But even if the search succeeded, it would still be improbable that the 'signal' would be a decodable message. It would more likely represent a by-product (or even a malfunction) of some super-complex machine far beyond our comprehension that could trace its lineage back to alien organic beings (which might still exist on their home planet or might long ago have died out). The only type of intelligence whose messages we could decode would be the (perhaps small) subset that used a technology attuned to our own parochial concepts. So, could we tell whether a signal is intended as a message or just some 'leakage'? Could we build up communication?

The philosopher Ludwig Wittgenstein said, 'If a lion could speak, we couldn't understand him'. Would the 'culture gap' with aliens be unbridgeable?

I don't think it necessarily would be. After all, if they managed to communicate, they would share with us an understanding of physics, mathematics, and astronomy. They may come from planet Zog and have seven tentacles; they may be metallic and electronic. But they would be made of similar atoms to us; they would (if they had eyes) stare out at the same cosmos and trace their origins back to the same hot dense beginning—the 'big bang' around 13.8 billion years ago. But there's no hope for snappy repartee—if they exist, they would be so far away that exchanging messages would take decades, or even centuries.

Even if intelligence were widespread in the cosmos, we may only ever recognise a small and atypical fraction of it. Some 'brains' may package reality in a fashion that we can't conceive. Others could be living contemplative energy-conserving lives, doing nothing to reveal their presence. It makes sense to focus searches first on Earthlike planets orbiting long-lived stars. But science fiction authors remind us that there are more exotic alternatives. In particular, the habit of referring to an 'alien civilisation' may be too restrictive. A 'civilisation' connotes a society of individuals; in contrast, ET might be a single integrated intelligence. Even if signals were being transmitted, we may not recognise them as artificial because we may not know how to decode them. A veteran radio engineer familiar only with amplitude modulation might have a hard time decoding modern wireless communications. Indeed, compression techniques aim to make the signal as close to noise as possible—insofar as a signal is predictable, there's scope for more compression.

The focus has been on the radio part of the spectrum. But of course, in our state of ignorance about what might be out there, we should explore all wavebands; we should look in the optical and X-ray band and also be alert for other evidence of nonnatural phenomena or activity. One might seek evidence for artificially created molecules such as CFCs in an exoplanet atmosphere, or else for massive artefacts such as a Dyson sphere. (This idea, due to Freeman Dyson, envisions that an energy-profligate civilisation might harness all the energy of its parent star by surrounding it with photovoltaic cells, and that the 'waste heat' would emerge as infrared emission.) And it's worth looking for artefacts within our solar system; maybe we can rule out visits by human-scale aliens, but if an extraterrestrial civilisation had mastered nanotechnology and transferred its intelligence to machines, the 'invasion' might consist of a swarm of microscopic probes that could have evaded notice. It's even worth keeping an eye open for especially shiny or oddly shaped objects lurking among the asteroids. But it would of course be easier to send a radio or laser signal than to traverse the mind-boggling distances of interstellar space.

I don't think even the optimistic SETI searchers would rate the chance of success as more than a few percent—and most of us are more pessimistic. But it's so fascinating that it seems worth a gamble— we'd all like to see searches begun in our lifetime. And there are two familiar maxims that pertain to this quest: 'Extraordinary claims will require extraordinary evidence', and, 'Absence of evidence isn't evidence of absence'.

Also, we have to realise just how surprising some natural phenomena can be. For instance, in 1967 Cambridge astronomers found regular radio 'beeps', repeating several times a second. Could this have been an alien transmission? Some were open to accepting this option, but soon it became clear that these beeps came from a hitherto undetected kind of very dense object: neutron stars, which are only a few kilometres across and spin at several revs per second (sometimes several hundred), sending a 'lighthouse beam' of radiation towards us from deep space. The study of neutron stars—of which thousands are now known—has proved an especially exciting and fruitful topic because they manifest extreme physics, where nature has created conditions that we could never simulate in the laboratory.[8] More recently, a new and still perplexing class of 'radio bursts' has been discovered, emitting even more powerfully than pulsars,[9] but the general disposition is to seek natural explanations for them.

SETI depends on private philanthropy. The failure to get public funds surprises me. If I were up before a government committee, I'd feel less vulnerable and more at ease defending a SETI project than seeking funds for a vast new particle accelerator. That's because many thousands of those watching movies of the Star Wars genre would be happy if some of the tax revenues they generated were hypothecated for SETI.

Perhaps we'll one day find evidence of alien intelligence—or even (though this is less likely) 'plug in' to some cosmic mind. On the other hand, our Earth may be unique and the searches may fail. This would disappoint the searchers. But it would have an upside for humanity's long-term resonance. Our solar system is barely middle-aged, and if humans avoid self-destruction within the next century, the posthuman era beckons. Intelligence from Earth could spread through the entire galaxy, evolving into a teeming complexity far beyond what we can even conceive. If so, our tiny planet—this pale blue dot floating in space—could be the most important place in the entire cosmos.

Either way, our cosmic habitat—this immense firmament of stars and galaxies—seems 'designed' or 'tuned' to be an abode for life. From a simple big bang, amazing complexity has unfolded, leading to our emergence. Even if we are now alone in the universe, we may not be the culmination of this 'drive' towards complexity and consciousness. This tells us something very profound about nature's laws—and motivates a brief excursion, in the following chapters, out to the broadest horizons in time and space that cosmologists conceive.

# 4

# THE LIMITS AND FUTURE OF SCIENCE

## 4.1. FROM THE SIMPLE TO THE COMPLEX

A fictional speculation: suppose a 'time machine' allowed us to send one succinct 'tweet' to great scientists of the past—Newton or Archimedes, for instance. What message would most enlighten them and transform their vision of the world? I think it would be the marvellous realisation that we ourselves, and everything in the everyday world, are made from fewer than one hundred different kinds of atoms—lots of hydrogen, oxygen, and carbon; small but crucial admixtures of iron, phosphorous, and other elements. All materials—living and nonliving—owe their structures to the intricate patterns in which atoms stick together, and how they react. The whole of chemistry is determined by the interactions between the positively charged nuclei of atoms and the negatively charged swarm of electrons that they're embedded in.

Atoms are simple; we can write down the equations of quantum mechanics (Schrödinger's equation) that describe their properties. So, on the cosmic scale, are black holes, for which we can solve Einstein's equations. These 'basics' are well enough understood to enable engineers to design all the objects of the modern world. (Einstein's theory of general relativity has found practical use in GPS satellites; their clocks would lose accuracy if they weren't properly corrected for the effects of gravity.)

The intricate structure of all living things testifies that layer on layer of complexity can emerge from the operation of underlying laws. Mathematical games can help to develop our awareness of how simple rules, reiterated over and over again, can indeed have surprisingly complex consequences.

John Conway, now at Princeton University, is one of the most charismatic figures in mathematics.[1] When he taught at Cambridge, students created a 'Conway appreciation society'. His academic research deals with a branch of mathematics known as group theory. But he reached a wider audience and achieved a greater intellectual impact through developing the Game of Life.

In 1970 Conway was experimenting with patterns on a Go board; he wanted to devise a game that would start with a simple pattern and use basic rules to iterate again and again. He discovered that by adjusting the rules of his game and the starting patterns, some arrangements produce incredibly complicated results—seemingly from nowhere because the rules of the game are so basic. 'Creatures' emerged, moving around the board, that seemed to have a life of their own. The simple rules merely specify when a white square turns into a black square (and vice versa), but, applied over and over again, a fascinating variety of complicated patterns is created. Devotees of the game identified objects such as 'glider', 'glider gun', and other reproducing patterns.

Conway indulged in a lot of 'trial and error' before he came up with a simple 'virtual world' that allowed for interesting emergent variety. He used pencil and paper, before the days of personal computers, but the implications of the Game of Life only emerged when the greater speed of computers could be harnessed. Likewise, early PCs enabled Benoit Mandelbrot and others to plot out the marvellous patterns of fractals—showing how simple mathematical formulas can encode intricate apparent complexity.

Most scientists resonate with the perplexity expressed in a classic essay by the physicist Eugene Wigner, titled 'The Unreasonable Effectiveness of Mathematics in the Natural Sciences'.[2] And also with Einstein's dictum that 'the most incomprehensible thing about the universe is that it is comprehensible'. We marvel that the physical world isn't anarchic— that atoms obey the same laws in distant galaxies as in our laboratories. As I've already noted (section 3.5), if we ever discover aliens and want to communicate with them, mathematics, physics, and astronomy would be perhaps the only shared culture. Mathematics is the language of science—and has been ever since the Babylonians devised their calendar and predicted eclipses. (Some of us would likewise regard music as the language of religion.)

Paul Dirac, one of the pioneers of quantum theory, showed how the internal logic of mathematics can point the way towards new discoveries. Dirac averred that 'the most powerful method of advance is to employ all the resources of pure mathematics in attempts to perfect and generalise the mathematical formalism that forms the existing basis of theoretical physics and—after each success in this direction—to try to interpret the new mathematical features in terms of physical entities'.[3] It was this approach—following the mathematics where it leads—that led Dirac to the idea of antimatter: 'antielectrons', now known as positrons, were discovered just a few years after he formulated an equation that would have seemed ugly without them.

Present-day theorists, with the same motives as Dirac, are hoping to understand reality at a deeper level by exploring concepts such as string theory, involving scales far smaller than any we can directly probe. Likewise, at the other extreme, some are exploring cosmological theories that offer intimations that the universe is vastly more extensive than the 'patch' we can observe with our telescopes (see section 4.3).

Every structure in the universe is composed of basic 'building blocks' governed by mathematical laws. However, the structures are generally too complicated for even the most powerful computers to calculate. But perhaps in the far-distant future, posthuman intelligence (not in organic form, but in autonomously evolving objects) will develop hypercomputers with the processing power to simulate living things—even entire worlds. Perhaps advanced beings could use hypercomputers to simulate a 'universe' that is not merely patterns on a chequerboard (like Conway's game) or even like the best 'special effects' in movies or computer games. Suppose they could simulate a universe fully as complex as the one we perceive ourselves to be in. A disconcerting thought (albeit a wild speculation) then arises: perhaps that's what we really are!

## 4.2. MAKING SENSE OF OUR COMPLEX WORLD

Possibilities once in the realms of science fiction have shifted into serious scientific debate. From the very first moments of the big bang to the possibilities for alien life, scientists are led to worlds even weirder than most fiction writers envision. At first sight one might think it presumptuous to claim—or even seek—to understand the remote cosmos when there's so much that baffles us closer at hand. But that's not necessarily a fair assessment. There is nothing paradoxical about the whole being simpler than its parts. Imagine an ordinary brick—its shape can be described in a few numbers. But if you shatter it, the fragments can't be described so succinctly.

Scientific progress seems patchy. Odd though it may seem, some of the best-understood phenomena are far away in the cosmos. Even in the seventeenth century, Newton could describe the 'clockwork of the heavens'; eclipses could be both understood and predicted. But few other things are so predictable, even when we understand them. For instance, it's hard to forecast, even a day before, whether those who travel to view an eclipse will encounter clouds or clear skies. Indeed, in most contexts, there's a fundamental limit to how far ahead we can predict. That's because tiny contingencies—like whether or not a butterfly flaps its wings—have consequences that grow exponentially. For reasons like this, even the most fine-grained computation cannot normally forecast British weather even a few days ahead. (But—and this is important—this doesn't stymie predictions of long-term climate change, nor weaken our confidence that it will be colder next January than it is in July.)

Today, astronomers can convincingly attribute tiny vibrations in a gravitational-wave detector to a 'crash' between two black holes more than a billion light years from Earth.[4] In contrast, our grasp of some familiar matters that interest us all—diet and child care, for instance—is still so meagre that 'expert' advice changes from year to year. When I was young, milk and eggs were thought to be good; a decade later they were deemed dangerous because of their high cholesterol content; and now they seem again to be deemed harmless. So lovers of chocolate and cheese may not have to wait long before being told those foods are good for them. And there is still no cure for many of the commonest ailments.

But it actually isn't paradoxical that we've achieved confident understanding of arcane and remote cosmic phenomena while being flummoxed by everyday things. It's because astronomy deals with phenomena far less complex than the biological and human sciences (even than 'local' environmental sciences).

\*   \*   \*

So how should we define or measure complexity? A formal definition was suggested by the Russian mathematician Andrey Kolmogorov: an object's complexity depends on the length of the shortest computer programme that could generate a full description of it.

Something made of only a few atoms cannot be very complicated. Big things need not be complex either. Consider, for instance, a crystal—even if it were large it wouldn't be called complex. The recipe for (for instance) a salt crystal is short: take sodium and chlorine atoms and pack them together, over and over again, to make a cubical lattice. Conversely, if you take a large crystal and chop it up, there is little change until it is broken down to the scale of single atoms. Despite its vastness, a star is fairly simple too. Its core is so hot that no chemicals can exist (complex molecules get torn apart); it is basically an amorphous gas of atomic nuclei and electrons. Indeed, black holes, exotic though they seem, are among the simplest entities in nature. They can be described precisely by equations no more complicated than those that describe a single atom.

Our high-tech objects are complex. For instance, a silicon chip with a billion transistors has structure on all levels down to a few atoms. But most complex of all are living things. An animal has interlinked internal structure on several different scales—from the proteins in single cells, right up to limbs and major organs. It doesn't preserve its essence if it is chopped up. It dies. Humans are more complex than atoms or stars (and, incidentally, midway between them in mass; it takes about as many human bodies to make up the Sun as there are atoms in each of us). The genetic recipe for a human being is encoded in three billion links of DNA. But we are not fully determined by our genes; we are moulded by our environment and experiences. The most complex things we know about in the universe are our own brains. Thoughts and memories (coded by neurons in the brain) are far more varied than genes.

There's an important difference, however, between 'Kolmogorov complexity' and whether something actually looks complicated. For instance, Conway's Game of Life leads to complicated-looking structures. But these can all be described by a short programme: take a particular starting position, and then iterate, over and over again, according to the simple rules of the game. The intricate fractal pattern of Mandelbrot's set is likewise the result of a simple algorithm. But these are exceptions. Most things in our everyday environment are too complicated to be predicted, or even fully described in detail. But much of their essence can nonetheless be captured by a few key insights. Our perspective has been transformed by great unifying ideas. The concept of continental drift (plate tectonics) helps us to fit together a whole raft of geological and ecological patterns across the globe. Darwin's insight—evolution via natural selection—reveals the overarching unity of the entire web of life on this planet. And the double helix of DNA reveals the universal basis for heredity. There are patterns in nature. There are even patterns in how we humans behave—in how cities grow, how epidemics spread, and how technologies like computer chips develop. The more we understand the world, the less bewildering it becomes and the more we're able to change it.

The sciences can be viewed as a hierarchy, ordered like the floors of a building, with those dealing with more complex systems higher up: particle physics in the basement, then the rest of physics, then chemistry, then cell biology, then botany and zoology, and then the behavioural and human sciences (with the economists claiming the penthouse).

The 'ordering' of the sciences in this hierarchy is not controversial. But what is more controversial is the sense in which the 'ground floor sciences'—particle physics in particular—are deeper or more fundamental than the others. In one sense they truly are. As the physicist Steven Weinberg has pointed out: 'The arrows all point downward'. Put another way, if you go on asking Why? Why? Why? you end

up at the particle level. Scientists are nearly all re-
ductionists in Weinberg's sense; they feel confident
that everything, however complex, is a solution of
Schrödinger's equation—unlike the 'vitalists' of earlier eras, who thought that living things were infused
with some special 'essence'. But this reductionism
isn't conceptually useful. As another great physicist,
Philip Anderson, emphasised, 'more is different';
macroscopic systems that contain large numbers of
particles manifest 'emergent' properties and are best
understood in terms of new concepts appropriate to
the level of the system.

Even a phenomenon as un-mysterious as the flow
of water in pipes or rivers is understood in terms of
'emergent' concepts like viscosity and turbulence.
Specialists in fluid mechanics don't care that water is
actually made up of $H_2O$ molecules; they see water
as a continuum. Even if they had a hypercomputer
that could solve Schrödinger's equation for the flow,
atom by atom, the resultant simulation wouldn't
provide any insight into how waves break, or what
makes a flow become turbulent. And new irreducible
concepts are even more crucial to our understanding of really complicated phenomena—for instance,
migrating birds or human brains. Phenomena on

different levels of the hierarchy are understood in terms of different concepts—turbulence, survival, alertness, and so forth. The brain is an assemblage of cells; a painting is an assemblage of pigments. But what is important and interesting is the pattern and structure—the emergent complexity.

That's why the analogy with a building is a poor one. The entire structure of a building is imperilled by weak foundations. In contrast, the 'higher level' sciences dealing with complex systems aren't vulnerable to an insecure base, as a building is. Each science has its own distinct concepts and modes of explanation. Reductionism is true in a sense. But it's seldom true in a *useful* sense. Only about 1 percent of scientists are particle physicists or cosmologists. The other 99 percent work on 'higher' levels of the hierarchy. They're challenged by the complexity of their subject—not by any deficiencies in our understanding of subnuclear physics.

## 4.3. HOW FAR DOES PHYSICAL REALITY EXTEND?

The Sun formed 4.5 billion years ago, but it's got around 6 billion years more before its fuel runs out. It will then flare up, engulfing the inner planets.

And the expanding universe will continue—perhaps forever—destined to become ever colder, ever emptier. To quote Woody Allen, eternity is very long, especially towards the end.

Any creatures witnessing the Sun's demise won't be human—they'll be as different from us as we are from a bug. Posthuman evolution—here on Earth and far beyond—could be as prolonged as the Darwinian evolution that has led to us—and even more wonderful. And evolution is now accelerating; it can happen via 'intelligent design' on a technological time-scale, operating far faster than natural selection and driven by advances in genetics and in artificial intelligence (AI). The long-term future probably lies with electronic rather than organic 'life' (see section 3.3).

In cosmological terms (or indeed in a Darwinian time frame) a millennium is but an instant. So let us 'fast forward' not for a few centuries, or even for a few millennia, but for an 'astronomical' timescale millions of times longer than that. The 'ecology' of stellar births and deaths in our galaxy will proceed gradually more slowly, until jolted by the 'environmental shock' of an impact with the Andromeda Galaxy, maybe four billion years hence. The debris of our galaxy, Andromeda, and their smaller companions—which now make up what is called the Local Group—will thereafter aggregate into one amorphous swarm of stars.

On the cosmic scale, gravitational attraction is overwhelmed by a mysterious force latent in empty space that pushes galaxies apart from each other. Galaxies accelerate away and disappear over a horizon—rather like an inside-out version of what happens when something falls into a black hole. All that will be left in view, after a hundred billion years, will be the dead and dying stars of our Local Group. But these could continue for trillions of years—time enough, perhaps, for the long-term trend for living systems to gain complexity and 'negative entropy' to reach a culmination. All the atoms that were once in stars and gas could be transformed into structures as intricate as a living organism or a silicon chip—but on a cosmic scale. Against the darkening background, protons may decay, dark matter particles annihilate, occasional flashes when black holes evaporate—and then silence.

In 1979, Freeman Dyson (already mentioned in section 2.1) published a now-classic article whose aim was 'to establish numerical bounds within which the universe's destiny must lie'.[5] Even if all material were optimally converted into a computer or superintelligence, would there still be limits on how much information could be processed? Could an unbounded number of thoughts be thought? The answer depends on the cosmology. It takes less energy to carry out computations at low temperatures. For the universe we seem to be in, Dyson's limit would be finite, but would be maximised if the 'thinkers' stayed cool and thought slowly.

Our knowledge of space and time is incomplete. Einstein's relativity (describing gravity and the cosmos) and the quantum principle (crucial for understanding the atomic scale) are the two pillars of twentieth-century physics, but a theory that unifies them is unfinished business. Current ideas suggest that progress will depend on fully understanding what might seem the simplest entity of all—'mere' empty space (the vacuum) is the arena for everything that happens; it may have a rich texture, but on scales a trillion trillion times smaller than an atom. According to string theory, each 'point' in ordinary space might, if viewed with this magnification, be revealed as a tightly folded origami in several extra dimensions.

The same fundamental laws apply throughout the entire domain we can survey with telescopes.

Were that not so—were atoms 'anarchic' in their behaviour—we'd have made no progress in understanding the observable universe. But this observable domain may not be all of physical reality; some cosmologists speculate that 'our' big bang wasn't the only one—that physical reality is grand enough to encompass an entire 'multiverse'.

We can only see a finite volume—a finite number of galaxies. That's essentially because there's a horizon, a shell around us, delineating the greatest distance from which light can reach us. But that shell has no more physical significance than the circle that delineates your horizon if you're in the middle of the ocean. Even conservative astronomers are confident that the volume of space-time within range of our telescopes—what astronomers have traditionally called 'the universe'—is only a tiny fraction of the aftermath of the big bang. We'd expect far more galaxies located beyond the horizon, unobservable, each of which (along with any intelligences it hosts) will evolve rather like our own.

It's a familiar idea that if enough monkeys were given enough time, they would write the works of Shakespeare (and indeed all other books, along with every conceivable string of gobbledygook).

This statement is mathematically correct. But the number of 'failures' that would precede eventual success is a number with about ten million digits. Even the number of atoms in the visible universe has only eighty digits. If all the planets in our galaxy were crawling with monkeys, who had been typing ever since the first planets formed, then the best they would have done is typed a single sonnet (their output would include short coherent stretches from all the world's literatures, but no single complete work). To produce a specific set of letters as long as a book is so immensely improbable that it wouldn't have happened even once within the observable universe. When we throw dice we eventually get a long succession of sixes, but (unless they are biased) we wouldn't expect to get more than a hundred in a row even if we went on for a billion years.

However, if the universe stretches far enough, everything could happen—somewhere far beyond our horizon there could even be a replica of Earth. This requires space to be VERY big—described by a number not merely with a million digits but with 10 to the power of 100 digits: a one followed by one hundred zeroes. Ten to the power of 100 is called a googol, and a number with a googol of zeros is a googolplex.

Given enough space and time, all conceivable chains of events could be played out somewhere, though almost all of these would occur far out of range of any observations we could conceivably make. The combinatorial options could encompass replicas of ourselves, taking all possible choices. Whenever a choice has to be made, one of the replicas will take each option. You may feel that a choice you make is 'determined'. But it may be a consolation that, somewhere far away (far beyond the horizon of our observations) you have an avatar who has made the opposite choice.

All this could be encompassed within the aftermath of 'our' big bang, which could extend over a stupendous volume. But that's not all. What we've traditionally called 'the universe'—the aftermath of 'our' big bang—may be just one island, just one patch of space and time, in a perhaps infinite archipelago. There may have been many big bangs, not just one. Each constituent of this 'multiverse' could have cooled down differently, maybe ending up governed by different laws. Just as Earth is a very special planet among zillions of others, so—on a far grander scale—could our big bang have been a rather special one. In this hugely expanded cosmic perspective, the laws of Einstein and the quantum could be mere parochial bylaws governing our cosmic patch. So, not only could space and time be intricately 'grainy' on a submicroscopic scale, but also, at the other extreme—on scales far larger than astronomers can probe—it may have a structure as intricate as the fauna of a rich ecosystem. Our current concept of physical reality could be as constricted, in relation to the whole, as the perspective of the Earth available to a plankton whose 'universe' is a spoonful of water.

Could this be true? A challenge for twenty-first-century physics is to answer two questions. First, are there many 'big bangs' rather than just one? Second—and this is even more interesting—if there are many, are they all governed by the same physics?

If we're in a multiverse, it would imply a fourth and grandest Copernican revolution; we've had the Copernican revolution itself, then the realisation that there are billions of planetary systems in our galaxy; then that there are billions of galaxies in our observable universe. But now that's not all. The entire panorama that astronomers can observe could be a tiny part of the aftermath of 'our' big bang, which is itself just one bang among a perhaps infinite ensemble.

(At first sight, the concept of parallel universes might seem too arcane to have any practical impact. But it may [in one of its variants] actually offer the prospect of an entirely new kind of computer: the quantum computer, which can transcend the limits of even the fastest digital processor by, in effect, sharing the computational burden among a near infinity of parallel universes.)

Fifty years ago, we weren't sure whether there had been a big bang. My Cambridge mentor Fred Hoyle, for instance, contested the concept, favouring a 'steady state' cosmos that was eternal and unchanging. (He was never fully converted—in his later years he espoused a compromise idea that might be called a 'steady bang'.) Now we have enough evidence to delineate cosmic history back to the ultradense first nanosecond—with as much confidence as a geologist inferring the early history of Earth. So in fifty more years, it is not overoptimistic to hope that we may have a 'unified' physical theory, corroborated by experiment and observation in the everyday world, that is broad enough to describe what happened in the first trillionth of a trillionth of a trillionth of a second—where the densities and energies were far higher than the range in which current theories apply. If that future theory were to predict multiple big bangs we should take that prediction seriously, even though it can't be directly verified (just as we give credence to what Einstein's theory tells us about the unobservable insides of black holes, because the theory has survived many tests in domains we can observe).

We may, by the end of this century, be able to ask whether or not we live in a multiverse, and how much variety its constituent 'universes' display. The answer to this question will determine how we should interpret the 'biofriendly' universe in which we live (sharing it with any aliens with whom we might one day make contact).

My 1997 book, *Before the Beginning*,[6] speculated about a multiverse. Its arguments were partly motivated by the seemingly 'biophilic' and fine-tuned character of our universe. This would occasion no surprise if physical reality embraced a whole ensemble of universes that 'ring the changes' on the basic constants and laws. Most would be stillborn or sterile, but we would find ourselves in one of those where the laws permitted emergent complexity. This idea had been bolstered by the 'cosmic inflation' theory of the 1980s, which offered new insights into how our entire observable universe could have 'sprouted' from an event of microscopic size. It gained further serious attention when string theorists began to favour the possibility of many different vacuums—each an arena for microphysics governed by different laws.

I've ever since had a close-up view of this shift in opinion and the emergence of these (admittedly speculative) ideas. In 2001, I helped organise a conference on this theme. It took place in Cambridge, but not in the university. I hosted it at my home, a farmhouse on the edge of the city, in a converted barn that offered a somewhat austere location for our discussions. Some years later, we had a follow-up conference. This time the location was very different: a rather grand room in Trinity College, with a portrait of Newton (the college's most famous alumnus) behind the podium.

The theorist Frank Wilczek (famous for his role, while still a student, in formulating what is called the 'standard model' of particle physics) attended both meetings. When he spoke at the second, he contrasted the atmosphere at the two gatherings.

He described physicists at the first meeting as 'fringe' voices in the wilderness who had for many years promoted strange arguments about conspiracies among fundamental constants and alternative universes. Their concerns and approaches seemed totally alien to the consensus vanguard of theoretical physics, which was busy successfully constructing a unique and mathematically perfect universe. But at the second meeting, he noted that 'the vanguard had marched off to join the prophets in the wilderness'.

Some years ago, I was on a panel at Stanford University where we were asked by the chairman: 'On the scale, "would you bet your goldfish, your dog, or your life," how confident are you about the multiverse concept?' I said that I was nearly at the dog level. Andrei Linde, a Russian cosmologist who had spent twenty-five years promoting a theory of 'eternal inflation' said he'd almost bet his life. Later, on being told this, the eminent theorist Steven Weinberg said he'd happily bet Martin Rees's dog and Andrei Linde's life.

Andrei Linde, my dog, and I will all be dead before this is settled. It's not metaphysics. It's highly speculative. But it's exciting science. And it may be true.

## 4.4. WILL SCIENCE 'HIT THE BUFFERS'?

A feature of science is that as the frontiers of our knowledge are extended, new mysteries, just beyond the frontiers, come into sharper focus. Unexpected discoveries have been perennially exciting in my own subject of astronomy. In every subject there will, at every stage, be 'unknown unknowns'. (Donald Rumsfeld was mocked for saying this in a different context—but of course he was right, and it might have been better for the world had he become a philosopher.) But there is a deeper question. Are there things that we'll never know, because they are beyond the power of human minds to grasp? Are our brains matched to an understanding of all key features of reality?

We should actually marvel at how much we have understood. Human intuition evolved to cope with the everyday phenomena our remote ancestors encountered on the African savanna. Our brains haven't changed much since that time, so it is remarkable that they can grasp the counterintuitive behaviours of the quantum world and the cosmos. I conjectured earlier that answers to many current mysteries will come into focus in the coming decades. But maybe not all of them; some key features of reality may be beyond our conceptual grasp. We may sometime 'hit the buffers'; there may be phenomena, crucial to our long-term destiny and to a full understanding of physical reality, that we are not aware of, any more than a monkey comprehends the nature of stars and galaxies. If aliens exist, some may have 'brains' that structure their consciousness in a fashion that we can't conceive and that have a quite different perception of reality.

We are already being aided by computational power. In the 'virtual world' inside a computer, astronomers can mimic galaxy formation, or crash another planet into the Earth to see if that's how the Moon might have formed; meteorologists can simulate the atmosphere, for weather forecasts and to predict long-term climatic trends; brain scientists can simulate how neurons interact. Just as video games get more elaborate as their consoles get more powerful, so, as computer power grows, these 'virtual' experiments become more realistic and useful.

Furthermore, there is no reason why computers can't actually make discoveries that have eluded unaided human brains. For example, some substances are perfect conductors of electricity when cooled to very low temperatures (superconductors). There is a continuing quest to find the 'recipe' for a superconductor that works at ordinary room temperatures (the highest superconducting temperature achieved so far is about −135 degrees Celsius at normal pressures and somewhat higher, about −70 degrees, for hydrogen sulphide at very high pressure). This would allow lossless transcontinental transmission of electricity, and efficient 'mag-lev' trains.

The quest involves a lot of 'trial and error'. But it's becoming possible to calculate the properties of materials, and to do this so fast that millions of alternatives can be computed, far more quickly than actual experiments could be performed. Suppose that a machine came up with a unique and successful recipe. It might have succeeded in the same way as AlphaGo. But it would have achieved something that would earn a scientist a Nobel prize. It would have behaved as though it had insight and imagination within its rather specialised universe—just as AlphaGo flummoxed and impressed human champions with some of its moves. Likewise, searches for the optimal chemical composition for new drugs will increasingly be done by computers rather than by real experiments, just as for many years aeronautical engineers have simulated air flow over wings by computer calculations rather than depending on wind-tunnel experiments.

Equally important is the capability to discern small trends or correlations by 'crunching' huge data sets. To take an example from genetics, qualities like intelligence and height are determined by combinations of genes. To identify these combinations would require a machine fast enough to scan large samples of genomes to identify small correlations. Similar procedures are used by financial traders in seeking out market trends and responding rapidly to them, so that their investors can top-slice funds from the rest of us.

My claim that there are limits to what human brains can understand was, incidentally, contested by David Deutsch, a physicist who has pioneered key concepts of 'quantum computing'. In his provocative and excellent book *The Beginning of Infinity*,[7] he pointed out that any process is in principle computable. This is true. However, being able to compute something is not the same as having an insightful comprehension of it. Consider an example from geometry, where points in the plane are designated by two numbers, the distance along the $x$-axis and along the $y$-axis. Anyone who has studied geometry at school would recognise the equation $x^2 + y^2 = 1$ as describing a circle. The famous Mandelbrot set is described by an algorithm that can be written down in a few lines. And its shape can be plotted by even a modestly powered computer—its 'Kolmogorov complexity' isn't high. But no human who is just given the algorithm can grasp and visualise this immensely complicated 'fractal' pattern in the same sense that they can visualise a circle.

We can expect further dramatic advances in the sciences during this century. Many questions that now perplex us will be answered, and new questions will be posed that we can't even conceive today. We should nonetheless be open-minded about the possibility that despite all our efforts, some fundamental truths about nature could be too complex for unaided human brains to fully grasp. Indeed, perhaps we'll never understand the mystery of these brains themselves—how atoms can assemble into 'grey matter' that can become aware of itself and ponder its origins. Or perhaps any universe complicated enough to have allowed our emergence is for just that reason too complicated for our minds to understand.

Whether the long-range future lies with organic posthumans or with intelligent machines is a matter for debate. But we would be too anthropocentric if we believed that a full understanding of physical reality is within humanity's grasp, and that no enigmas will remain to challenge our posthuman descendants.

## 4.5. WHAT ABOUT GOD?

If the number one question astronomers are asked is, Are we alone?, the number two question is surely, Do you believe in God? My conciliatory answer is that I do not, but that I share a sense of wonder and mystery with many who do.

The interface between science and religion still engenders controversy, even though there has been no essential change since the seventeenth century. Newton's discoveries triggered a range of religious (and antireligious) responses. So, even more, did Charles Darwin in the nineteenth century. Today's scientists evince a variety of religious attitudes; there are traditional believers as well as hard-line atheists among them. My personal view—a boring one for those who wish to promote constructive dialogue (or even just unconstructive debate) between science and religion—is that, if we learn anything from the pursuit of science, it is that even something as basic as an atom is quite hard to understand. This should induce scepticism about any dogma, or any claim to have achieved more than a very incomplete and metaphorical insight into any profound aspect of existence. As Darwin said, in a letter to the American biologist Asa Gray: 'I feel most deeply that the whole subject is too profound for the human intellect. A dog might as well speculate on the mind of Newton. Let each man hope and believe as he can'.[8]

Creationists believe that God created the Earth more or less as it is—leaving no scope for emergence of new species or enhanced complexity and paying little regard to the wider cosmos. It is impossible to refute, by pure logic, even someone who claims that the universe was created an hour ago, along with all our memories and all vestiges of earlier history. 'Creationist' concepts still hold sway among many US evangelicals and in parts of the Muslim world. In Kentucky there is a 'creation museum' with what its promoters describe as a 'full-size' Noah's Ark, 510 feet long, built at a cost of $150 million.

A more sophisticated variant—'intelligent design'—is now more fashionable. This concept accepts evolution but denies that random natural selection can account for the immensely long chain of events that led to our emergence. Much is made of stages where a key component of living things seems to have required a series of evolutionary steps rather than a single leap, but where the intermediate steps would in themselves confer no survival advantage. But this style of argument is akin to traditional creationism. The 'believer' focuses on some details (and there are many) that are not yet understood and argues that the seeming mystery constitutes a fundamental flaw in the theory. Anything can be 'explained' by invoking supernatural intervention. So, if success is measured by having an explanation, however 'flip', then the 'intelligent designers' will always win.

But an explanation only has value insofar as it integrates disparate phenomena and relates them to a single underlying principle or unified idea. Such a principle is Darwinian natural selection as expounded in *On the Origin of Species*, a book he described as 'one long argument'. Actually, the first great unifying idea was Newton's law of gravity, identifying the familiar gravitational pull that holds us on the ground and makes an apple fall with the force that holds the Moon and planets in their orbits. Because of Newton, we need not record the fall of every apple.

Intelligent design dates back to classic arguments: a design needs a designer. Two centuries ago, the theologian William Paley introduced the now-well-known metaphor of the watch and the watchmaker—adducing the eye, the opposable thumb, and so forth as evidence of a benign Creator.[9] We now view any biological contrivance as the outcome of prolonged evolutionary selection and symbiosis with its surroundings. Paley's arguments have fallen from favour even among theologians.[10]

Paley's view of astronomy was that it was not the most fruitful science for yielding evidence of design, but 'that being proved, it shows, above all others, the scale of [the Creator's] operations'. Paley might have reacted differently if he'd known about the providential-seeming physics that led to galaxies, stars, planets, and the distinctive elements of the periodic table. The universe evolved from a simple beginning—a 'big bang'—specified by quite a short recipe. But the physical laws are 'given' rather than having evolved. Claims that this recipe seems rather special can't be so readily dismissed as Paley's biological 'evidences' (and a possible explanation in terms of a multiverse is mentioned in section 4.3).

A modern counterpart of Paley, the ex-mathematical physicist John Polkinghorne, interprets our fine-tuned habitat as 'the creation of a Creator who wills that it should be so'.[11] I have had genial public debates with Polkinghorne; he taught me physics when I was a Cambridge student. The line I take is that his theology is too anthropocentric and constricted to be credible. He doesn't espouse 'intelligent design' but believes that God can influence the world by giving a nudge or tweak at places and times when the outcome is especially responsive to small changes—maximum impact with a minimal and readily concealed effort.

When meeting Christian clergy (or their counterparts in other faiths), I try to enquire about what they consider the 'bottom line'—'the theoretical minimum' that must be accepted by their adherents. It's clear that many Christians regard the resurrection as a historical and physical event. Polkinghorne certainly does; he dresses it up as physics, saying that Christ transitioned to an exotic material state that will befall the rest of us when the apocalypse comes. And in his 2018 Easter message, the Archbishop of Canterbury, Justin Welby, said that if the resurrection is 'just a story or metaphor, frankly, I should resign from my job'. But how many Catholics really believe in the two miracles—the 'practical' part of the examination—that a potential candidate must achieve in order to qualify for sainthood? I'm genuinely perplexed that so many have a faith that has such literal content.

I would describe myself as a practising but unbelieving Christian. The parallel concept is familiar among Jews: there are many who follow traditional observances—lighting candles on Friday nights and so forth. But this need not mean that they accord their religion any primacy, still less that they claim it has any unique truth. They may even describe themselves as atheists. Likewise, as a 'cultural Christian', I'm content to participate (albeit irregularly) in the rituals of the Anglican church with which I've been familiar since early childhood.

Hard-line atheists focus too much, however, on religious dogma and on what is called 'natural theology'—seeking evidence of the supernatural in the physical world. They must be aware of 'religious' people who are manifestly neither unintelligent nor naive. By attacking mainstream religion, rather than striving for peaceful coexistence with it, they weaken the alliance against fundamentalism and fanaticism. They also weaken science. If a young Muslim or evangelical Christian is told that they can't have their God and accept evolution, they will opt for their God and be lost to science. Adherents of most religions accord high importance to their faith's communal and ritual aspects—indeed many of them might prioritise ritual over belief. When so much divides us, and change is disturbingly fast, such shared ritual offers bonding within a community. And religious traditions, linking adherents with past generations, should strengthen our concern that we should not leave a degraded world for generations to come.

This line of thought segues into my final theme: how should we respond to the challenges of the twenty-first century and narrow the gap between the world as it is and the world we'd like to live in and share with the rest of 'creation'?

# 5

# CONCLUSIONS

## 5.1. DOING SCIENCE

Chapter 1 of this book highlighted the transformations occurring this century—unprecedented in their speed and in the stress they impose on the global environment. Chapter 2 focused on the scientific advances that we can expect in the coming decades, emphasising the benefits but also the ethical dilemmas and the risk of disruption or even catastrophe. Chapter 3 explored broader horizons in both space and time—speculating about domains far beyond our planet, and the prospects for a 'posthuman' future. Chapter 4 assessed the prospects of understanding ourselves and the world more deeply—what we may learn, and what may forever be beyond our grasp. In the last few pages I focus closer to the here and now—to explore, against this backdrop, the role of scientists. I distinguish their special obligations from those that fall to all of us, as humans and as citizens concerned about the world future generations will inherit.

But first, an important clarification: I'm using 'science' throughout as a shorthand that embraces technology and engineering as well. Harnessing and implementing a scientific concept for practical goals can be a greater challenge than the initial discovery. A favourite cartoon of my engineering friends shows two beavers looking up at a vast hydroelectric dam. One beaver says to the other: 'I didn't actually build it, but it's based on my idea'. And I like to remind my theorist colleagues that the Swedish engineer Gideon Sundback, who invented the zipper, made a bigger intellectual leap than most of us ever will.

Scientists are widely believed to follow a distinctive procedure that's described as the scientific method. This belief should be laid to rest. It would be truer to say that scientists follow the same rational style of reasoning as lawyers or detectives in categorising phenomena and assessing evidence. A related (and indeed damaging) misperception is the widespread presumption that there is something especially 'elite' about the quality of their thought. 'Academic ability' is one facet of the far wider concept of intellectual ability—possessed in equal measure by the best journalists, lawyers, engineers, and politicians. E. O. Wilson (the ecologist quoted in section 1.4) avers that to be effective in some scientific fields it's actually best not to be too bright.[1] He's not disparaging the insights and eureka moments that punctuate (albeit rarely) scientists' working lives. But, as the world expert on tens of thousands of ant species, Wilson's research has involved decades of hard slog: armchair theorising is not enough. So, there is a risk of boredom. And he is indeed right that those with short attention spans—with 'grasshopper minds'—may find happier (albeit less worthwhile) employment as 'millisecond traders' on Wall Street, or the like.

Scientists generally pay too little regard to philosophy, but some philosophers resonate with them. Karl Popper, in particular, caught the fancy of scientists in the second half of the twentieth century.[2] He was right to say that a scientific theory must be in principle refutable. If a theory is so flexible—or its proponents so seemingly shifty—that it can be adjusted to fit any eventuality, then it isn't genuine science. Reincarnation is an example. In a well-known book, the biologist Peter Medawar[3] somewhat more controversially berated Freudian psychoanalysis on this score, putting the knife in firmly at the end: 'Considered in its entirety, psychoanalysis won't do. It is an end product, moreover, like a dinosaur or a zeppelin, no better theory can ever be erected on its ruins, which will remain for ever one of the saddest and strangest of all landmarks in the history of twentieth century thought'. But the Popper doctrine nonetheless has two weaknesses. Firstly, interpretation depends on the context. Consider, for instance, the Michelson-Morley experiment, which showed, at the end of the nineteenth century, that the speed of light (measured by a clock in the laboratory) was the same however fast the laboratory was moving—and the same at all times of the year, despite the Earth's motion. This was later realised to be a natural consequence of Einstein's theory. But had the same experiment been performed in the seventeenth century, it would have been adduced as evidence that the Earth didn't move—and claimed as a refutation of Copernicus. Secondly, judgment is needed in deciding how compelling the contrary evidence needs to be before a well-supported theory is abandoned. As Francis Crick, codiscoverer of the structure of DNA, reputedly said, if a theory agrees with all the facts it is bad news, because some 'facts' are likely to be wrong.

Second only to Popper, it is the American philosopher Thomas Kuhn—with his concept of 'normal science' being punctuated by 'paradigm shifts'—who has gained traction.[4] The Copernican revolution, overturning the concept of an Earth-centred cosmos, qualifies as a paradigm shift. So does the realisation that atoms are governed by quantum effects—utterly counterintuitive and still mysterious. But many of Kuhn's disciples (though maybe not Kuhn himself) used the phrase too freely. For instance, it's routinely claimed that Einstein overthrew Newton. But it's fairer to say that he transcended Newton. Einstein's theory applied more widely—to contexts where forces are very strong or speeds very high—and gave a much deeper understanding of gravity, space, and time. Piecemeal modification of theories, and their absorption in new ones of greater generality, has been the pattern in most sciences.[5]

The sciences demand a range of different types of expertise and different styles; they can be pursued by speculative theorists, by lone experimenters, by ecologists gaining data in the field, and by quasi-industrial teams working on giant particle accelerators or big space projects. Most commonly, scientific work involves collaboration and debate in a small research group. Some people aspire to write a pioneering paper opening up a subject; others gain more satisfaction from writing a definitive monograph tidying up and codifying a topic after it's become well understood.

Indeed, the sciences are as diverse as sports. It's hard for generic writing about sports to get beyond vacuous generalities—extolling humanity's competitive streak and so forth. It's more interesting to write about the distinctive features of a particular sport; still more compelling are the particularities of especially exciting games and the personalities of key players. So it is with the sciences. Each particular science has its methods and conventions. And what most compels our interest is the fascination of the individual discovery or insight.

The cumulative advance of science requires new technology and new instruments—in symbiosis, of course, with theory and insight. Some instruments are 'tabletop' in scale. At the other extreme, the Large Hadron Collider at CERN in Geneva, 9 kilometres in diameter, is currently the world's most elaborate scientific instrument. Its completion in 2009 generated enthusiastic razzmatazz and wide public interest, but at the same time questions were understandably raised about why such a large investment was being made in the seemingly recondite science of subnuclear physics. But what is special about this branch of science is that its practitioners in many different countries have chosen to commit most of their resources over a time-span of nearly twenty years to construct and operate a single vast instrument in a European-led collaboration. The annual contribution made by participant nations (like the United Kingdom) amounts to only about 2 percent of their overall budget for academic science, which doesn't seem a disproportionate allocation to a field so challenging and fundamental. This global collaboration on a single project to probe some of nature's most fundamental mysteries—and push technology to its limits—is surely something in which our civilisation can take pride. Likewise, astronomical instruments are run by multinational consortia, and some are truly global projects—for instance, the ALMA radio telescope in Chile (Atacama Large Millimeter/Submillimeter Array) has participation from Europe, the United States, and Japan.

Those embarking on research should pick a topic to suit their personality, and also their skills and tastes (for fieldwork? For computer modelling? For high-precision experiments? For handling huge data sets? And so forth). Moreover, young researchers can expect to find it especially gratifying to enter a field where things are advancing fast—where you have access to novel techniques, more powerful computers, or bigger data sets—so that the experience of the older generation is at a deep discount. And another thing: it is unwise to head straight for the most important or fundamental problem. You should multiply the importance of the problem by the probability that you'll solve it, and maximise that product. Aspiring scientists shouldn't all swarm into, for instance, the unification of cosmos and quantum, even though it's plainly one of the intellectual peaks we aspire to reach, and they should realise that the great challenges in cancer research and in brain science need to be tackled in a piecemeal fashion, rather than head-on. (As mentioned in section 3.5, investigating the origin of life used to be in this category, but it now is deemed timely and tractable in a way it wasn't until recently.)

What about those who switch to a new field of science in mid-career? The ability to bring in new insights, and a new perspective, is a 'plus'—indeed, the most vibrant fields often cut across traditional disciplinary boundaries, On the other hand, it's conventional wisdom that scientists don't improve with age—that they 'burn out'. The physicist Wolfgang Pauli had a famous put-down for scientists past thirty: 'still so young, and already so unknown'. But I hope it's not just wishful thinking on the part of an aging scientist to be less fatalistic. There seem to be three destinies for us. First, and most common, is a diminishing focus on research—sometimes compensated by energetic efforts in other directions, sometimes just by a decline into torpor. A second pathway, followed by some of the greatest scientists, is an unwise and overconfident diversification into other fields. Those who follow this route are still, in their own eyes, 'doing science'—they want to understand the world and the cosmos, but they no longer get satisfaction from researching in the traditional piecemeal way: they over-reach themselves, sometimes to the embarrassment of their admirers. This syndrome has been aggravated by the tendency for the eminent and elderly to be shielded from criticism—though one of the many benefits of a less hierarchical society is that this insulation is now rarer, at least in the West; moreover, the increasingly collaborative nature of science makes isolation less likely. But there is a third way—the most admirable. This is to continue to do what one is competent at, accepting that there may be some new techniques that the young can assimilate more easily than the old, and that one can probably at best aspire to be on a plateau rather than scaling new heights.

There are some 'late-flowering' exceptions. But whereas there are many composers whose last works are their greatest, there are few scientists for whom this is so. The reason, I think, is that composers, though influenced in their youth (like scientists) by the then-prevailing culture and style, can thereafter improve and deepen solely through 'internal development'. Scientists, in contrast, need continually to absorb new concepts and new techniques if they want to stay at the frontier—and that's what gets harder as we get older.

Many sciences—astronomy and cosmology among them—advance decade by decade so that practitioners can observe an 'arc of progress' during their career. Paul Dirac, a leader in the extraordinary revolution in the 1920s that codified quantum theory, said that it was an era when 'second-rate' people did 'first-rate' work. Luckily for my generation of astronomers, that's been true in our field in recent decades.

The best laboratories, like the best start-ups, should be optimal incubators of original ideas and young talent. But there's an insidious demographic trend that militates against this in traditional universities and institutes. Fifty years ago, the science profession was still growing exponentially, riding on the expansion of higher education, and the young outnumbered the old; moreover, it was normal (and generally mandatory) to retire by one's mid-sixties. The academic community, at least in the West, isn't now expanding much (and in many areas has reached saturation level), and there is no enforced retirement age. In earlier decades, it was reasonable to aspire to lead a group by one's early thirties—but in the United States' biomedical community, it's now unusual to get your first research grant before the age of forty. This is a very bad augury. Science will always attract 'nerds' who can't envisage any other career. And laboratories can be staffed with those content writing grant applications, which usually fail to get funding. But the profession needs to attract a share of those with flexible talent, and the ambition to achieve something by their thirties. If that perceived prospect evaporates, such people will shun academia, and maybe attempt a start-up. This route offers great satisfaction and public benefit— many should take it—but in the long run it's important that some such people dedicate themselves to the fundamental frontiers. The advances in IT and computing can be traced back to basic research done in leading universities—in some cases nearly a century ago. And the stumbling blocks encountered in medical research stem from uncertain fundamentals. For instance, the failure of anti-Alzheimer's drugs to pass clinical tests, which has caused Pfizer to abandon its programme to develop neurological drugs, may indicate that not enough is known about how the brain functions and that the effort should refocus on basic science.

The expansion of wealth and leisure—coupled with the connectivity offered by IT—will offer millions of highly educated amateurs and 'citizen scientists' anywhere in the world greater scope than ever before to follow their interests. Moreover, these trends will enable leading researchers to do cutting-edge work outside a traditional academic or governmental laboratory. If enough make this choice, it will erode the primacy of research universities and boost the importance of 'independent scientists' to the level that prevailed before the twentieth century— and perhaps enhance the flowering of genuinely original ideas.

## 5.2. SCIENCE IN SOCIETY

A main theme of this book is that our future depends on making wise choices about key societal challenges: energy, health, food, robotics, environment, space, and so forth. These choices involve science. But key decisions shouldn't be made just by scientists; they matter to us all and should be the outcome of wide public debate. For that to happen, we all need enough 'feel' for the key ideas of science, and enough numeracy to assess hazards, probabilities, and risks, so as not to be bamboozled by experts or credulous of populist sloganising.

Those who aspire to a more engaged democracy routinely bemoan how little the typical voter knows about the relevant issues. But ignorance isn't peculiar to science. It's equally sad if citizens don't know their nation's history, can't speak a second language, and can't find North Korea or Syria on a map—and many can't. (In one survey, only a third of Americans could find Britain!) This is an indictment of our education system and culture in general—I don't think scientists have a special reason to moan. Indeed, I'm gratified and surprised that so many people are interested in dinosaurs, Saturn's moons, and the Higgs boson—all blazingly irrelevant to our day-to-day lives—and that these topics feature so frequently in the popular media.

Moreover, quite apart from their practical use, these ideas should be part of our common culture. More than that, science is the one culture that's truly global: protons, proteins, and Pythagoras are the same from China to Peru. Science should transcend all barriers of nationality. And it should straddle all faiths too. It's an intellectual deprivation not to understand our natural environment and the principles that govern the biosphere and climate. And to be blind to the marvellous vision offered by Darwinism and modern cosmology—the chain of emergent complexity leading from a 'big bang' to stars, planets, biospheres, and human brains—rendering the cosmos aware of itself. These 'laws' or patterns are the great triumphs of science. To discover them required dedicated talent—even genius. And great inventions require equivalent talent. But to grasp the key ideas isn't so difficult. Most of us appreciate music even if we can't compose it, or even perform it. Likewise, the key ideas of science can be accessed and enjoyed by almost everyone—if conveyed using nontechnical words and simple images. The technicalities may be daunting, but they can be left to the specialists.

Advances in technology have led to a world where most people enjoy a safer, longer, and more satisfying life than previous generations, and these positive trends could continue. On the other hand, environmental degradation, unchecked climate change, and unintended downsides of advanced technology are collaterals of these advances. A world with a higher population more demanding of energy and resources, and more empowered by technology, could trigger serious, even catastrophic, setbacks to our society.

The public is still in denial about two kinds of threats: harm that we're causing collectively to the biosphere, and threats that stem from the greater vulnerability of our interconnected world to error or terror induced by individuals or small groups. Moreover, what's new in this century is that a catastrophe will resonate globally. In his book *Collapse*,[6]

Jared Diamond describes how and why five different societies have decayed or encountered catastrophes and gives contrasting prognoses for some modern societies. But these events weren't global; for instance, the Black Death didn't reach Australia. But in our networked world, there would be nowhere to hide from the consequences of economic collapse, a pandemic, or a collapse in global food supplies. And there are other global threats; for instance, intense fires after a nuclear exchange could create a persistent 'nuclear winter'—preventing, in worst-case scenarios, the growing of conventional crops for several years (as could also happen after an asteroid impact or a super-volcano eruption).

In such a predicament it is collective intelligence that would be crucial. No single person fully understands the smartphone—a synthesis of several technologies. Indeed, if we were stranded after an 'apocalypse', as in extreme survival movies, even the basic technologies of the iron age and agriculture would be beyond almost all of us. That's why, incidentally, James Lovelock—the polymath who introduced the Gaia hypothesis (the self-regulating planetary ecology)—has urged that 'handbooks for survival', codifying basic technology, should be prepared, widely dispersed, and securely stored. This challenge has been taken up, for instance, by the UK astronomer Lewis Dartnell in his excellent book *The Knowledge: How to Rebuild Our World from Scratch.*[7]

More should be done to assess, and then minimise, the probability of global hazards. We live under their shadow, and they are raising the stakes for humanity. The emergent threat from technically empowered mavericks is growing. The issues impel us to plan internationally (for instance, whether or not a pandemic gets global grip may hinge on how quickly a Vietnamese poultry farmer can report any strange sickness). And many of the challenges—for instance, planning how to meet the world's energy needs while avoiding dangerous climate change, and ensuring food-source security for nine billion people without jeopardising a sustainable environment—involve multidecade timescales that are plainly far outside the 'comfort zone' of most politicians. There's an institutional failure to plan long-term and to plan globally.

There's no denying that futuristic technologies, if misapplied, could lead to hazards, even catastrophes. It is important to take advantage of the best expertise to assess which risks are credible, and which can be dismissed as science fiction, and to focus precautionary measures on the former. How can this be done? It's not feasible to control the rate of advance, still less to relinquish potentially hazardous developments completely, unless a single organisation holds the purse strings—and that's completely unrealistic in a globalised world with a mix of commercial, philanthropic, and governmental funding. But even if regulations can't be anywhere near 100 percent effective—and can provide little more than a 'nudge'— it's important for the scientific community to do all it can to promote 'responsible innovation'. In particular, it may be crucial to influence the order in which various innovations come to fruition. For instance, if a superpowerful AI 'went rogue', it would then be too late to control other developments; on the other hand, an AI firmly under human control, but highly accomplished, could aid in reducing the risk from biotech or nanotechnology.

Nations may need to give up more sovereignty to new global organisations along the lines of the International Atomic Energy Agency, the World Health Organization, and so on. There are already international bodies that regulate air travel, radio frequency allocations, and such. And there are protocols such as the post-Paris climate change agreement. More such bodies may be needed, for planning energy generation, to ensure sharing of water resources, and for responsible exploitation of AI and of space technology. National boundaries are now being eroded, not least by the quasi-monopolies like Google and Facebook. The new organisations must retain accountability to governments but will need to use social media—as they are now and will be in future decades—and involve the public. Social media engage huge numbers in campaigns, but the barrier to their engagement is so low that most lack the commitment of participants in mass movements in the past. Moreover, the media make it easy to engineer a protest, as well as amplifying all dissident minorities—adding to the challenge of governance.

But will the world be governable by nation-states? Two trends are reducing interpersonal trust: firstly, the remoteness and globalisation of those we routinely have to deal with; and secondly, the rising vulnerability of modern life to disruption—the realisation that 'hackers' or dissidents can trigger incidents that cascade globally. Such trends necessitate burgeoning security measures. These are already irritants in our everyday life—security guards, knotty passwords, airport searches, and so forth—but they are likely to become ever more vexatious. Innovations like blockchain, the publicly distributed ledger that combines open access with security, could offer protocols that render the entire internet more secure. But their current applications—allowing an economy based on crypto-currencies to function independently of traditional financial institutions—seem damaging rather than benign. It's both salutary and depressing to realise how much of the economy is dedicated to activities and products that would be superfluous if we felt we could trust each other.

The gaps in wealth and welfare levels between countries show little sign of narrowing. But if they persist, the risk of persistent disruption will grow. This is because the disadvantaged are aware of the injustice of their predicament; travel is easier, and therefore more aggressive measures will be needed in order to control migratory pressures if they build up. But apart from direct transfers of funds in the traditional way, the internet and its successors should make it easier for services to be provided anywhere in the world, and for educational and health benefits to spread more widely. It's in the interests of the wealthy world to invest massively in improving the quality of life and job opportunities in poorer countries—minimising grievances and 'levelling up' the world.

## 5.3. SHARED HOPES AND FEARS

All scientists have special obligations over and above their responsibility as citizens. There are ethical obligations confronting scientific research itself: avoiding experiments that have even the tiniest risk of leading to catastrophe and respecting a code of ethics when research involves animals or human subjects. But less tractable issues arise when their research has ramifications beyond the laboratory and has a potential social, economic, and ethical impact that concerns all citizens—or when it reveals a serious but still-unappreciated threat. You would be a poor parent if you didn't care what happened to your children in adulthood, even though you may have little control over them. Likewise, scientists shouldn't be indifferent to the fruits of their ideas—their creations. They should try to foster benign spin-offs—commercial or otherwise. They should resist, so far as they can, dubious or threatening applications of their work, and alert politicians when appropriate. If their findings raise ethical sensitivities—as will happen acutely and often—they should engage with the public, while realising that they have no distinct credentials outside their specialism.

One can highlight some fine exemplars from the past: for instance, the atomic scientists who developed the first nuclear weapons during World War II. Fate had assigned them a pivotal role in history. Many of them—men such as Joseph Rotblat, Hans Bethe, Rudolf Peierls, and John Simpson (all of whom I was privileged to know in their later years)—returned with relief to peacetime academic pursuits. But for them the ivory tower wasn't a sanctuary. They continued not just as academics but as engaged citizens—promoting efforts to control the power they had helped unleash, through national academies, the Pugwash movement, and other public forums. They were the alchemists of their time, possessors of secret specialised knowledge.

The technologies I've discussed in earlier chapters have implications just as momentous as nuclear weapons. But in contrast to the 'atomic scientists', those engaged with the new challenges span almost all the sciences, are broadly international—and work in the commercial sector as well as in academia and government. Their findings and concerns need to inform planning and policy. So how is this best done?

Direct ties forged with politicians and senior officials can help—and links with NGOs and the private sector too. But experts who've served as government advisors have often had frustratingly little influence. Politicians are, however, influenced by their inboxes, and by the press. Scientists can sometimes achieve more as 'outsiders' and activists, leveraging their message via widely read books, campaigning groups, blogging and journalism, or—albeit via a variety of perspectives—through political activity. If their voices are echoed and amplified by a wide public, and by the media, long-term global causes will rise on the political agenda.

Rachel Carson and Carl Sagan, for instance, were both preeminent in their generation as exemplars of the concerned scientist—and they had immense influence through their writings and speeches. And that was before the age of social media and tweets. Sagan, had he been alive today, would have been a leader of the 'marches for science'—electrifying crowds through his passion and eloquence.

A special obligation lies on those in academia or on self-employed entrepreneurs; they have more

freedom to engage in public debate than those em-
ployed in government service or in industry. Academics, moreover, have the special opportunity to
influence students. Polls show, unsurprisingly, that
younger people, who expect to survive most of the
century, are more engaged and anxious about long-
term and global issues. Student involvement in, for
instance, 'effective altruism' campaigns is burgeoning. William MacAskill's book *Doing Good Better*[8]
is a compelling manifesto. It reminds us that urgent
and meaningful improvements to people's lives can
be achieved by well-targeted redeployment of existing resources towards developing or destitute nations. Wealthy foundations have more traction (the
archetype being the Bill & Melinda Gates Foundation, which has had a massive impact, especially on
children's health)—but even they cannot match the
impact that national governments could have if there
were pressure from their citizens.

I've already highlighted the role of the world's
religions—transnational communities that think
long-term and care about the global community,
especially the world's poor. An initiative of a secular organisation, the California-based Long Now
Foundation, will create a symbol that contrasts

dramatically with our currently pervasive short-termism. In a cavern deep underground in Nevada, a massive clock will be built; it is designed to tick (very slowly) for ten thousand years, programmed to resound with a different chime every day over that expanse of time. Those of us who visit it in this century will contemplate a monument built to outlast the cathedrals, and will be inspired to hope that a hundred centuries from now it will indeed still be ticking—and that some of our progeny will still visit.

Although we live under the shadow of unfamiliar and potentially catastrophic hazards, there seems to be no *scientific* impediment to achieving a sustainable and secure world, where all enjoy a lifestyle better than those in the 'West' do today. We can be technological optimists, even though the balance of effort in technology needs redirection. Risks can be minimised by a culture of 'responsible innovation', especially in fields like biotech, advanced AI, and geoengineering, and by reprioritising the thrust of the world's technological effort. We should remain upbeat about science and technology—we shouldn't put the brakes on progress. Doctrinaire application of the 'precautionary principle' has a manifest downside. Coping with global threats requires more technology—but guided by social science and ethics.

The intractable geopolitics and sociology— the gap between potentialities and what actually happens—engenders pessimism. The scenarios I've described—environmental degradation, unchecked climate change, and unintended consequences of advanced technology—could trigger serious, even catastrophic, setbacks to society. But they have to be tackled internationally. And there's an institutional failure to plan for the long term, and to plan globally. Politicians look to their own voters—and the next election. Stockholders expect a payoff in the short run. We downplay what's happening even now in faraway countries. And we discount too heavily the problems we'll leave for new generations. Without a broader perspective—without realising that we're all on this crowded world together—governments won't properly prioritise projects that are long-term in a political perspective, even if a mere instant in the history of the planet.

'Space-Ship Earth' is hurtling through the void. Its passengers are anxious and fractious. Their life-support system is vulnerable to disruption and breakdowns. But there is too little planning, too little horizon scanning, too little awareness of long-term risks. It would be shameful if we bequeathed to future generations a depleted and hazardous world.

I began this book by quoting H. G. Wells. I end by recalling the words of a scientific sage from the second half of the last century, Peter Medawar: 'The bells which toll for mankind are—most of them anyway—like the bells on Alpine cattle; they are attached to our own necks, and it must be our fault if they do not make a cheerful and harmonious sound'.[9]

Now is the time for an optimistic vision of life's destiny—in this world, and perhaps far beyond it. We need to think globally, we need to think rationally, we need to think long-term—empowered by twenty-first-century technology but guided by values that science alone can't provide.

# NOTES

## CHAPTER 1. DEEP IN THE ANTHROPOCENE

1. The Earl of Birkenhead, *The World in 2030 AD* (London: Hodder and Stoughton, 1930).
2. Martin Rees, *Our Final Century* (London: Random House, 2003). The US edition (published by Basic Books) was retitled *Our Final Hour*.
3. H. G. Wells's lecture, 'The Discovery of the Future', was given at the Royal Institution, London, on January 24, 1902, and subsequently was published in a book with that title.
4. 'Resilient Military Systems and the Cyber Threat', Defense Science Board Report January 2013. Similar concerns have been reiterated by General Petraeus and other senior US figures.
5. The 2017 revision of the UN 'World Population Prospects' quotes a best estimate of 9.7 billion for the 2050 population. Another authoritative source is the Population Project of the International Institute for Applied Systems Analysis (IIASA), which estimates somewhat lower figures.
6. There are many reports on world food and water supplies—for instance, the 2013 report 'Modelling Earth's Future', jointly prepared by the Royal Society and the National Academy of Sciences.
7. 'Our Common Future', Report from the UN World Commission on Environment and Development, 1987.

8. Juncker's remark is quoted in the *Economist*, March 15, 2007.

9. The 'planetary boundaries' concept was spelled out in a 2009 report from the Stockholm Resilience Centre.

10. This quote is from E. O. Wilson's *The Creation: An Appeal to Save Life on Earth* (New York: W. W. Norton, 2006).

11. The conference, on May 2–6, 2014, was titled 'Sustainable Humanity, Sustainable Nature: Our Responsibility', and was cosponsored by the Pontifical Academy of Sciences and the Pontifical Academy of Social Sciences.

12. The quote is from Alfred Russel Wallace, *The Malay Archipelago* (London: Harper, 1869).

13. *The Skeptical Environmentalist* was published by Cambridge University Press in 2001. The Copenhagen Consensus, founded in 2002, is under the auspices of the Environmental Assessment Institute in Copenhagen.

14. The scientists involved in this project include C. Kennel at the University of California–San Diego, in La Jolla, and Emily Shuckburgh and Stephen Briggs in the United Kingdom.

15. The Stern Review Report on Economics of Climate Change, HM Treasury, UK, 2006.

16. G. Wagner and M. Woltzman, *Climate Shock and the Economic Consequences of a Hotter Planet* (Princeton, NJ: Princeton University Press, 2015).

17. W. Mischel, Y. Shoda, and M. L. Rodriguez, 'Delay of Gratification in Children', *Science* 244 (1989): 933–38.

18. 'Cuba's 100-Year Plan for Climate Change', *Science* 359 (2018): 144–45.

19. In the United Kingdom the case for the circular economy has gained traction through the advocacy of a widely admired high-profile figure, the around-the-world sailor Ellen MacArthur.

20. An excellent survey of geoengineering is Oliver Morton, *The Planet Remade: How Geoengineering Could Change the World* (Princeton: NJ: Princeton University Press, 2016).

## CHAPTER 2. HUMANITY'S FUTURE ON EARTH

1. Robert Boyle's archives, and this document in particular, are discussed by Felicity Henderson in a 2010 Royal Society Report.

2. This list can be found online at https://www.telegraph.co.uk /news/uknews/7798201/Robert-Boyles-Wish-list.html.

3. Two highly accessible books on these developments are Jennifer A. Doudna and Samuel S. Sternberg, *A Crack in Creation* (Boston: Houghton Mifflin Harcourt, 2017) (Jennifer Doudna is one of the inventors of CRISPR/Cas9); and Siddhartha Mukherjee, *The Gene: An Intimate History* (New York: Scribner, 2016).

4. The paper, by D. Evans and R. Noyce of the University of Alberta, is in *PLOS One* and is discussed in *Science News* on January 19, 2018. Ryan S. Noyce, Seth Lederman, and David H. Evans, 'Construction of an Infectious Horsepox Virus Vaccine from Chemically Synthesized DNA Fragments', *PLOS One* (January 19, 2018): https://doi.org/10.1371/journal .pone.0188453.

5. Chris D. Thomas, *Inheritors of the Earth* (London: Allen Lane, 2017).

6. Steven Pinker, *The Better Angels of Our Nature: Why Violence Has Declined* (New York: Penguin Books, 2011).

7. Freeman Dyson, *Dreams of Earth and Sky* (New York: Penguin Random House, 2015).

8. An overview of these developments is given in Murray Shanahan, *The Technological Singularity* (Cambridge, MA: MIT Press, 2015); and Margaret Boden, *AI: Its Nature and Future* (Oxford: Oxford University Press, 2016). A more speculative 'take' is offered by Max Tegmark, *Life 3.0: Being Human in the Age of Artificial Intelligence* (New York: Penguin Random House 2017).

9. David Silver et al., 'Mastering the Game of Go without Human Knowledge', *Nature* 550 (2017): 354–59.

10. https://en.wikipedia.org/wiki/Reported_Road_Casualties_Great_Britain.

11. The letter was organised by the Future of Life Institute, based at MIT.

12. Stuart Russell is quoted from the *Financial Times*, January 6, 2018.

13. See Ray Kurzweil, *The Singularity Is Near: When Humans Transcend Biology* (New York: Viking, 2005).

14. P. Hut and M. Rees, "How Stable Is Our Vacuum?" *Nature* 302 (1983): 508–9.

15. Derek Parfit's arguments are presented in part 4 of his *Reasons and Persons* (New York: Oxford University Press, 1984).

16. Good surveys of these extreme risks are given in Nick Bostrom and Milan Ćirković, eds., *Global Catastrophic Risks* (Oxford: Oxford University Press, 2011); and Phil Torres, *Morality, Foresight, and Human Flourishing: An Introduction to Existential Risks* (Durham, NC: Pitchstone, 2018).

## CHAPTER 3. HUMANITY IN A COSMIC PERSPECTIVE

1. Quoted in Carl Sagan, *Pale Blue Dot: A Vision of a Human Future in Space* (New York: Random House, 1994).

2. Alfred Russel Wallace, *Man's Place in the Universe* (London: Chapman and Hall, 1902)—this book can be downloaded free via the Gutenberg project.

3. Michel Mayor and Didier Queloz, 'A Jupiter-Mass Companion to a Solar-Type Star', *Nature* 378 (1995): 355–59.

4. The best information about the results from the *Kepler* spacecraft can be found on NASA's website: https://www.nasa.gov/mission_pages/kepler/main/index.html.

5. Michaël Gillon et al., 'Seven Temperate Terrestrial Planets Around the Nearby Ultracool Dwarf Star TRAPPIST-1', *Nature* 542 (2017): 456–60.

6. Laser acceleration was discussed by the visionary engineer Robert Forward in the 1970s. More recently, there have been detailed studies by P. Lubin, J. Benford, and others. And the Starshot Project, supported by Yuri Milner's Breakthrough Foundation, is seriously studying whether a wafer-sized probe could be accelerated to 20 percent of the speed of light, thereby reaching the nearest stars within twenty years.

7. Good introductions to the topic are Jim Al-Khalili, ed., *Aliens: The World's Leading Scientists on the Search for Extraterrestrial Life* (New York: Picador, 2017); and Nick Lane, *The Vital Question: Why Is Life the Way It Is?* (New York: W. W. Norton, 2015).

8. There is a huge literature on pulsars, but an overview is given by Geoff McNamara, *Clocks in the Sky: The Story of Pulsars* (New York: Springer, 2008).

9. Fast radio bursts are intensively studied and ideas are fast changing. The best reference is Wikipedia, https://en.wikipedia.org/wiki/Fast_radio_burst.

## CHAPTER 4. THE LIMITS AND FUTURE OF SCIENCE

1. A biography of Conway is Siobhan Roberts, *Genius at Play: The Curious Mind of John Horton Conway* (New York: Bloomsbury, 2015).

2. This essay can be found in Eugene Wigner, *Symmetries and Reflections: Scientific Essays of Eugene P. Wigner* (Bloomington: Indiana University Press, 1967).

3. The quote is from a classic 1931 paper by Paul Dirac titled 'Quantised Singularities in the Electromagnetic Field', *Proceedings of the Royal Society A*, 133 (1931): 60.

4. An excellent account of this discovery and its context is given by Govert Schilling in *Ripples in Spacetime* (Cambridge, MA: Belknap Press of Harvard University Press, 2017).

5. Freeman Dyson, 'Time without End: Physics and Biology in an Open Universe', *Reviews of Modern Physics* 51 (1979): 447–60.

6. Martin Rees, *Before the Beginning: Our Universe and Others* (New York, Basic Books, 1997).

7. David Deutsch, *The Beginning of Infinity: Explanations That Transform the World* (New York: Viking, 2011).

8. Darwin in a letter to Asa Gray written on May 22, 1860. Darwin Correspondence Project, Cambridge University Library.

9. William Paley, *Evidences of Christianity* (1802).

10. Parts of this section first appeared in Martin J. Rees, "Cosmology and the Mulitverse, in *Universe or Multiverse*, ed. Bernard Carr (Cambridge: Cambridge University Press, 2007).

11. John Polkinghorne, *Science and Theology* (London: SPCK/Fortress Press, 1995).

## CHAPTER 5. CONCLUSIONS

1. E. O. Wilson, *Letters to a Young Scientist* (New York: Liveright, 2014).

2. Karl Popper's key work on the scientific method is *The Logic of Scientific Discovery* (London: Routledge, 1959)—a translation of the original German version published in 1934. In the intervening years, Popper enhanced his reputation with his deeply impressive contribution to political theory: *The Open Society and Its Enemies.*

3. P. Medawar, *The Hope of Progress* (Garden City, NY: Anchor Press, 1973), 69.

4. T. S. Kuhn, *The Structure of Scientific Revolutions* (Chicago: University of Chicago Press, 1962).

5. The accessible book *The Meaning of Science*, by Tim Lewens (New York: Basic Books, 2016), offers a clear critique of the viewpoints of Popper, Kuhn, and others.

6. Jared Diamond, *Collapse: How Societies Choose to Fail or Succeed* (New York: Penguin, 2005).

7. Lewis Dartnell, *The Knowledge: How to Rebuild Our World from Scratch* (New York: Penguin, 2015). Books such as this are educative. It's surely regrettable that so many of us are ignorant of the basic technologies we depend on.

8. William MacAskill, *Doing Good Better: Effective Altruism and How You Can Make a Difference* (New York: Random House, 2016).

9. *The Future of Man* (1959).

# INDEX